The Green Design and Print Production Handbook

The Green Design and Print Production Handbook

**ADRIAN BULLOCK
& MEREDITH WALSH**

ILEX

THE GREEN DESIGN AND PRINT PRODUCTION HANDBOOK

First published in the United Kingdom in 2013 by
ILEX
210 High Street
Lewes
East Sussex
BN7 2NS

Distributed worldwide (except North America)
by Thames & Hudson Ltd.,
181A High Holborn, London WC1V 7QX,
United Kingdom

Copyright © 2013 The Ilex Press Limited

Publisher: Alastair Campbell
Creative Director: James Hollywell
Managing Editor: Nick Jones
Senior Editor: Ellie Wilson
Commissioning Editor: Emma Shackleton
Art Director: Julie Weir
Designer: Simon Goggin

Illustrations: Lucy Irving
http://www.lucyirving.com

British Library Cataloguing-in-Publication Data
A catalogue record for this book is available from the British Library.

ISBN: 978-1-78157-029-6

Colour Origination by Ivy Press Reprographics

Printed and bound in Hong Kong

This book was printed at an ISO 14001 accredited printer
on recycled wood free stock using vegetable-based inks.

10 9 8 7 6 5 4 3 2 1

Contents

Introduction

This book is aimed at anyone who has anything to do with print—from those in the publishing and printing industries to booksellers, reporters, authors, self-publishers, and writing groups. In short, this book is for anyone who wants to understand how what they do can and does affect the environment they inhabit and wants to find out what they can do about this in practical terms.

For many of us, the environment and what is happening to it seems so immense and so distant that as individuals we feel almost powerless to do anything about it. Our own efforts at recycling and reusing, at cutting down on travel and energy use, may make us feel good, but in reality these are quixotic when compared with the size and seriousness of the situation. We feel that the only real solution is going to come about through concerted inter-government action on a global basis.

In the meantime, climate change summits and conferences—Rio, Kyoto, Copenhagen, Durban—come and go, protocols are signed and then repealed, regulations and laws are passed and ignored, and all the while debate continues to rage over the possible causes of the environmental degradation of the world we live in and what should be done about it.

The intention behind this book is to turn that feeling of alienation, of powerlessness, into one in which you feel not only connected with what is happening and why, but you understand the practical part you can play in reducing your impact on the environment through your role as a publisher, printer, or author.

This book will provide you with an overview of what is happening now in order to give you an understanding of the situation as it stands and the issues. It will then move on to an exploration of possible options for changing the way we do things, before ending with a view of where we go from here.

Chapter 1: The Global Context considers the context in which the publishing industry is working as it seeks to create environmentally responsible business practices. The environment is changing; we look at why these changes are occurring and how they will affect the publishing industry. We find out more about the rapidly developing global drivers for change, the key players involved, the legislation, and voluntary codes. We look at the frameworks for action, which will help you achieve best practice in your own organization.

Chapter 2: The Environment and the Publishing Sector looks at the unique environmental issues the publishing industry faces, examining how it has responded to the need to reduce its impact on the environment, and the tools that have been created and adopted to achieve this.

The focus of the book is on creating and using processes that have a minimal impact on the environment, and in **Chapter 3:** Bottom Up and Top Down, we look at the value of putting all your business practices on a more environmentally sound footing. This chapter shows you how to change the culture of your business and make green thinking part of your daily routine, and it uncovers the unexpected business benefits this can provide. We show you how to set up an effective Environmental Management System to manage your green projects and the process of change.

In Chapter 4: Choosing and Setting up an Eco-Friendly Workflow, we start to develop the practical aspect of the book and explore the range of workflows available, recognizing that although it is difficult to find one size that fits all, there is a green supply chain workflow whose principles of dematerialization, detoxification, and de-energization (or de-carbonization) are designed to help you move towards producing and delivering products that are eco-friendly and sustainable.

The practical aspect continues in **Chapter 5:** How to Ensure You Buy Green Raw Materials, in which we pick our way through the maze of different papers, inks, and glues you use when you make your products. Which inks, papers, and glues really are eco-friendly? What exactly is recycled paper, and how, by using it, does it make your products more eco-friendly? Why is it that certain inks cause more air pollution than others?

Chapter 6: Integrating Green Practices into Prepress takes you through the process of planning green products. It starts at the planning and prepress stage when the inputs that define the outputs in terms of format, extent, paper, binding, and the use of color are decided, and ends by showing you the extent to which your decisions affect the eco-friendliness of your products, and how you can ensure that your products start green and end up green.

Working our way through the supply chain, we arrive at **Chapter 7:** Eco-Friendly Printing and Binding. This is the stage at which all the thinking and planning that have gone into creating your green product come together and are turned into printed and bound stock. We provide you with checklists so that you can work out for yourself how green your printers are and choose those who are most capable of producing your work in eco-friendly ways. You will also be able to consider different models for sourcing your printing and binding, not only technologically such as print on demand and digital printing, but also geographically through distributed printing.

By Chapter 8: Cutting back on Transport and Travel, we have reached the end of the supply chain, and we look at options for how, by reducing our transport and travel, we can reduce our environmental footprint. Digital technology helps us chase low manufacturing prices around the globe, and bringing stock back to base merely to repack it and send it off on its travels again isn't the only way of doing things. The same technology can be used in smart ways that make it possible for us to reduce the environmental impact of moving stock round the globe and to eliminate travel altogether.

In **Chapter 9**: Beyond the Door, we look at the responsibilities of the publisher once the product has left their direct control. What are the big issues for retailers, and how can publishers, printers, and retailers work together to address these, including the elephant in the room, "returns"? What happens to your products at the end of their life, and what options and opportunities are there to dispose of them environmentally? Are readers a sleeping giant who will wake to demand more from you, and, if so, what engagement with end users is needed now?

Chapter 10: The Future Is To Be Written. Here we look to the future, at the challenges for the publishing sector in a changing economic and environmental climate, and as it adopts new technologies. How can you make sure you are prepared to address them?

In writing this book, we have worked on, sifted through, and checked enormous amounts of information–fact as well as opinion and myth–to produce what we see as an authoritative and up-to-date guide to the current issues and challenges and how you can respond to them.

But we see our book as more than that; we see it as providing a vision for a workable and sustainable future in which you can play your part and make a difference. This may not happen immediately, and things will change with the passage of time. But you do need something to take you from where you are to where you want to be, and we hope that this book will fulfill this need.

Adrian Bullock and Meredith Walsh

The global context

In this chapter we will be looking at the global context within which we work. Why are we trying to make our publications greener? How do our actions affect the environment? In a time of rapid industry change for publishing and for the broader forest products industry, who is driving that change: is it governments, industry, or non-governmental organizations (NGOs), or a combination of all three, and how can we understand the changes they are making and incorporate these structures or lessons into our businesses? We will start by examining the stark realities of the environmental pressure our world is under, and then go on to look at how various groups and initiatives are trying to change this for the better.

Why should you care?

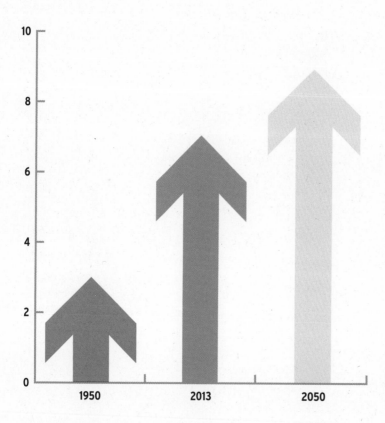

Left: Population growth in billions.

Every day we are confronted by evidence, in the press and in our working lives, that there is increasing pressure on the world's natural resources; this pressure on resources comes from population increases and high levels of consumption.

In 2011 the world population reached seven billion, and it is predicted to reach nine billion by 2050. This is a steep increase when you consider that in 1950 the global population had not yet reached three billion.[1] At the same time that we are seeing a population boom, we are also seeing changes to the standards and modes of living in large groups of people. An increasingly industrialized China and India are adopting a Western pattern of consumption that includes more books and printed materials, more electronic goods, and more meat in the diet, changing agricultural practice and putting pressure on land use. Populations in the West continue their habits of high consumption. These resource demands are having substantial effects on the world's forests, water availability, and carbon levels.

1. http://www.guardian.co.uk/environment/interactive/2011/oct/28/world-population-growth-7-billionth-person?INTCMP=SRCH

1   Population increase

2   Increase in levels of consumption

3   Increase in levels of consumption

4   Increase in pressure on resources

Above: Increases in consumption and
population have led to increased resource
use, with substantial effects on the world's
forests, water, and carbon levels.

Deforestation: biodiversity loss & climate change

Deforestation is occurring on an unprecedented scale and is predicted to continue on that trend given the increases in population and consumption.

You may largely hear of deforestation in the context of the decreasing habitat available to tigers or other large mammals, but deforestation is causing much larger-scale biodiversity loss in astonishingly biologically rich areas. This is a terrible loss, for science, for pharmaceutical discoveries, and simply for the plant and animal life that make up those forests as ecosystems. As an example of this extraordinary biodiversity, Ecuadorian scientists estimate it would take 400 years to map all the plant and animal species of the Yasuni region of the Amazon alone.[2]

Deforestation is a substantial contributor to climate change as well. The statistics in the World Resource Institute (WRI) report "Navigating the Numbers" speak volumes. Forestry and land-use change accounts for 18% of global greenhouse gases (GHGs); that's 4% more than transport, including aviation. And of the forestry GHG emissions, 34% are from Indonesia and 18% from Brazil.[3]

What happens to the forests of Indonesia has become an issue for everyone, as we will all see the effect of climate change. It is also the responsibility of many, as the products of the forests are sold globally; that pulp from these trees gets into paper and books sold in the U.S. and UK is now well documented.

Climate change, through weather disruptions, is responsible for the two recent dry spells in the Amazon: the first in 2005 was believed to be a once-in-a-lifetime

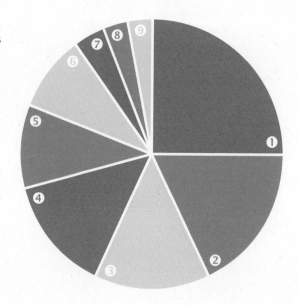

①	Electricity and heat 25%	⑥	Other fuel combustion 9%
②	Deforestation 18%	⑦	Fugitive emissions 4%
③	Transportation 14%	⑧	Waste 3%
④	Agriculture 14%	⑨	Industrial processes 3%
⑤	Industry energy use 10%		

Above: Global greenhouse gas emissions.

2. http://www.guardian.co.uk/environment/2011/dec/30/ecuador-paid-rainforest-oil-alliance?INTCMP=SRCH
3. http://pdf.wri.org/navigating_numbers.pdf

Right: Some climate models predict a dangerous feedback loop.

Below Right: Frogs are just one family of animals affected as deforestation causes large-scale biodiversity loss.

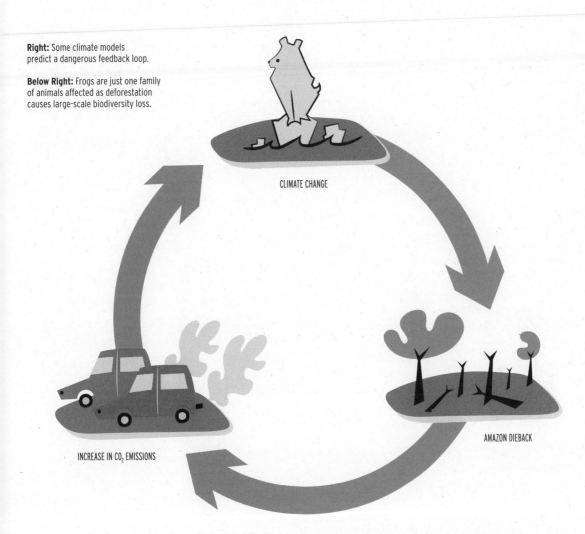

CLIMATE CHANGE

AMAZON DIEBACK

INCREASE IN CO_2 EMISSIONS

event but was followed by another severe dry spell in 2010. These two periods of drought have created significant dieback of the forest that in itself becomes a very significant source of CO_2. Some climate models have even suggested the Amazon could become a substantial source of CO_2 emissions, on a scale to interfere with any stabilization achieved elsewhere.[4] This in turn suggests a dangerous feedback loop of climate change causing Amazon dieback, which creates further climate change through CO_2 emissions.

4. http://assets.wwf.org.uk/downloads/tipping_point_report.pdf

Water shortages

Clean water is vital. It's essential for drinking and washing in the home, it's vital to agriculture, it's a key resource for many businesses, particularly manufacturing, and it is becoming scarce. Growing populations and consumption are increasing the requirement for water, and at the same time we are seeing disruptions to the water supply from drought, flooding, and pollution. The CEO of Nestlé has said that he believes we could run out of water before we run out of oil, and that water shortages in the next couple of decades could reduce global cereal production and trigger social unrest.[5]

Many manufactured products have a high embedded water footprint: for instance, the water footprint of a liter of milk is 264 gallons (1,000 liters), and the water footprint of 2.2 pounds (1 kilo) of cotton is 2,642 gallons (10,000 liters).[6] These are large volumes of water that are potentially not returned to the original water source; they are only partially returned or returned containing pollutants. Manufacturers use water in different ways, of course, and it is up to the different industries and individual companies to use water in the most effective way they can to reduce their water footprint. This could mean ensuring that either they take less water from the water table or that all the water that is returned to its original source is done so without pollutants. Paper manufacturing requires a lot of water (water is added to the pulp mix and the paper is formed during a swift-drying process), though the majority of this water can be captured and returned to its original water source. However, the manufacturer must make sure the water returned is chemical-free.

Changing weather patterns and land-use change are also creating water shortages. The floods in Pakistan, Brazil, and Australia between 2010 and 2011 all created drinking water shortages in those regions, and the frequency of such weather events seems to be increasing. But we are seeing droughts as well. The deforestation of the Amazon is thought to affect the rate at which the soil evaporates water, causing the area to become drier,[7] and this is at the same time as more water is required in the region because forest is being converted to farmland and water will be needed for crop growing.

When we consider the effects of land-use change along with the requirements of agriculture for water, we start to see how the effect of one action–deforestation– begins to fan out and affect much more than the immediate surroundings. The Water Resources Group (WRG) has highlighted the water-food-energy nexus where the interconnectedness between these three elements can be seen.[8] Any impact on one element will affect the others, and the net effect will be a combination of increased prices for everyone and further scarcity.

". . . food production requires water and energy; water extraction and distribution requires energy; and energy production requires water. Food prices are also highly sensitive to the cost of energy inputs through fertilizers, irrigation, transport, and processing." [9]

5. http://www.nestle.com/Media/NewsAndFeatures/Pages/Nestle-water-World-Economic-Forum.aspx
6. http://www.waterfootprint.org/?page=files/home
7. http://www.sciencedaily.com/releases/2011/06/110628111842.htm and http://wwf.panda.org/about_our_earth/about_forests/deforestation/

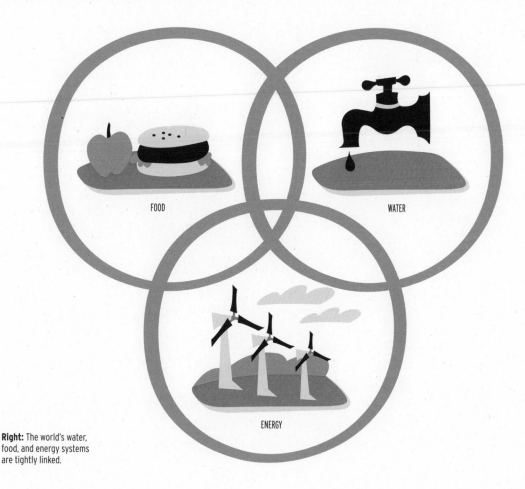

FOOD

WATER

ENERGY

Right: The world's water, food, and energy systems are tightly linked.

The picture we see emerging is of water as a communal resource that must be managed carefully as such through responsibly managed land use and from clean and efficient industry.

What we have described here are the interacting systems of forests, climate, water, and the changes created by human activity that we are seeing in them. These changes are the cause of cataclysmic events for the communities they affect, and they are occurring more often with more people affected directly through the loss of their homes, their livelihoods, and their water supply. There are also indirect effects. As communities are disrupted, supply chains are also disrupted and the costs of energy and goods are all going up. For our businesses and our communities to be sustainable, we must change our behavior, and there are organizations and initiatives that are aiming to do just that. In the next section, we look at the global drivers for change through legislation, voluntary codes, and frameworks for action.

There are many initiatives aimed at changing how human activity impacts on the world; those written about here have been chosen because they fall into one or more of the following categories:

• Legislation applicable to publishers or their supply chain
• Initiatives that set standards in mapping, and reporting on, enviromental impacts
• Initiatives that set a standard for performance and act as a tool to enable business to achieve change
• Initiatives that expand the current understanding of best practice

Collectively the initiatives discussed provide the global context we are working in.

8. http://www.weforum.org/videos/risks-focus-3-water-food-energy-nexus
9. ibid.

Forestry

Left: Each wooden piece of this guitar must be from legal origins to meet the Lacey Act requirements.

Forestry, as we have seen previously, plays an important role in climate change globally, in water availability locally, and in biodiversity loss. Outlined below are major initiatives that aim to stop illegal logging and instead move towards sustainable forestry. It's worth mentioning the order in which these initiatives are discussed.

Legislation is mentioned first as these legislative acts are likely to affect the majority of publishers. But the certification bodies were created ten or more years before the legislation, which shows what a vital role the NGOs and forestry industry have played, and will continue to play, in the move to sustainable forest management.

LACEY ACT

The Lacey Act in the U.S. was created in 1900 as a conservation law aimed at protecting wildlife by prohibiting the trade in illegally captured animals. In 2008 it was extended to cover the trade in and import of forest products. The law requires that all wood and wood-derived products, such as paper:

• shall be from legal sources
• that the company responsible for that import, i.e. the trader, not the shipping company, shall have taken due care to ensure that the material imported is from legal sources
• that the company responsible for the import shall be liable if any declaration is proved to be false.

The Act is enforced jointly by customs, the U.S. Fish and Wildlife Service, and the Animal Plant Health Inspection Service, reflecting the environmentally protective aim of the Act and the emphasis on import and trade.

There is currently no standardized declaration form, so there are some areas of contention about this Act. What constitutes "due care"? And how will it be proven that material is from legal sources? There are no absolute answers to these questions at the moment, but it is instructive to note some details in the case brought against Gibson guitars for the import of illegal ebony from India. The declaration supplied from India stated that the import was legal, as the ebony fretboards were finished. (Indian law allows for the export of finished ebony products but not unfinished ones: the law aims for both conservation and ensuring the best economic return for India from its resources.) However, the U.S. authorities asserted that, as the fretboards were blank, they were therefore unfinished.[10] The U.S. authorities took the declaration to be a false one. The Lacey Act, then, should be regarded as a conservation law that aims to stop illegal imports regardless of how the exporting country is enforcing its own law; and where there is ambiguity, the clearest solutions should be sought out. If the product imported is complex, then due diligence in its simplest, clearest form would be to ensure either that the country imported from is low risk for that particular material, or that the material imported is third-party certified. We will discuss certification shortly.

10. http://www.illegal-logging.info/uploads/ArnoldPorterLLPBNAWhiteCollarCrimeReport121611.pdf

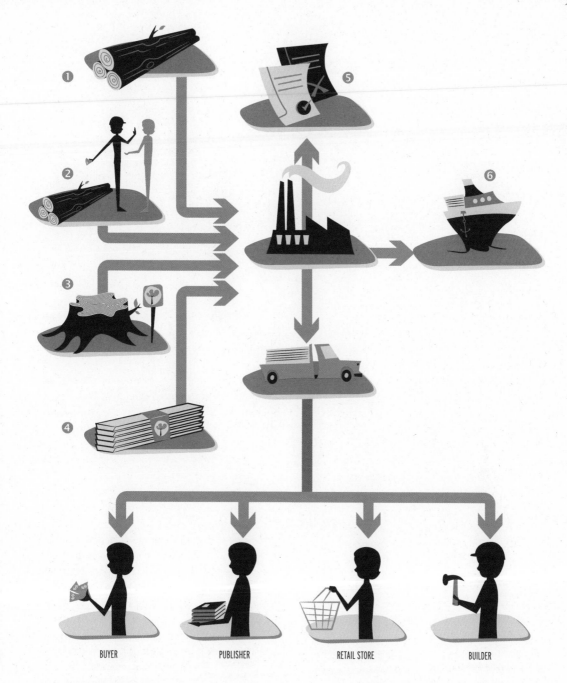

BUYER

PUBLISHER

RETAIL STORE

BUILDER

1 Logs cut from native lands without permission

2 Failure to pay statutory tariffs at forest concession

3 Old growth trees cut from National Park

4 Timber labeled as another species

5 Counterfeit permits issued

6 Logs shipped against log-export ban

Source: International Centre for Trade and Sustainable Development/The Environmental Investigation Agency

EU FLEGT AND TIMBER REGULATION

The European Union (EU) has adopted a two-pronged approach to stopping the trade of illegal timber with its FLEGT (Forest Law Enforcement, Governance, and Trade) action plan: these are the Voluntary Partnership Agreements (VPAs) and the EU Timber Regulation (EUTR). The EU is creating bilateral VPAs between the producer countries and the EU. These partnership agreements recognize the consumer countries' essential role in engaging with the producers and taking responsibility in helping to stop the illegal trade of timber. The VPAs will do the following:

• Create a licensing system under which legal products can be identified
• Prevent the import of unlicensed products
• Build capacity to set up the licensing system in the producer country
• Build capacity to improve enforcement of forest laws
• Build capacity to reform laws where necessary

The EUTR focuses on the domestic laws of the EU, and on the importers of forest products coming from any country that does not have a VPA in place. It will do the following:

• Examine the EU member states' domestic legislation
• Consider adding to the domestic legislation
• Require a due diligence system of importers, either their own or a monitoring organization
• Prohibit the placing on the market of products either illegally imported into or illegally produced in the EU

The first two points aim to ensure there is no weak point of entry into the EU through which illegal imports could be brought in. Once within the EU those forest products will be out of the scope of the EU Timber Regulation as this regulation applies to products only when they are first placed into the EU market. The second two points place responsibility with the importing company to assess their supply chain and remove any illegally imported timber from it.

This law was agreed in 2010 and will come into force in March 2013. It is important to note that printed material is currently excluded.

Printed material was a late exclusion from the regulation, and this is likely to have been because of the complexity of paper as a product. Paper can have as many as twenty different pulp sources, all of which would need to have gone through a process of due diligence. Publishers, who are ultimately responsible for printed

Right: Forest Stewardship Council (FSC)
"tick tree" logo.

Below Right: Example of an FSC crate label.

Part No.			
36	**175**	**2.5**	
		0204XX	FSC
Number of Pieces **400**		Date **08/11/12**	
		Species **SPRUCE**	
Ilex Press Ltd. Tel: 01273 403124 Fax: 01273 487441			

material, are several steps removed from the forest
sources, and would need to create a robust due diligence
system that can risk assess all pulp sources and therefore
the paper being used (more on this in Chapter 2, pp. 40–41).
This means that when the regulation is first introduced
in 2013, the pulps imported into and manufactured in
Europe will be covered by the regulation, but pulps in
printed books imported into Europe will not be. But as
certification and monitoring bodies respond to this
challenge, as they are sure to do—and with due diligence
systems encompassing certified and non-certified sources
already existing—it is likely that printed material will not
remain excluded.

CERTIFICATION SYSTEMS

One of the answers to deforestation is to put in place
systems of responsible forestry that aim to acknowledge
and protect biodiversity, land rights of indigenous peoples,
sacred spaces, and the commercial and social needs of
the local community. In the 1990s the two most widely
recognized certification systems were created with the
aim of doing just that.

FSC AND PEFC

The Forest Stewardship Council (FSC) was set up after the
1992 Rio UN Summit on Sustainable Development, which
had highlighted the need for responsible forestry because
of the effects of forestry on the world's ecosystems.

In particular, the FSC's aim was to halt the sale of
illegal tropical hardwoods for furniture into the UK by
creating a system that would allow those responsible
for manufacturing to trace material and state accurately
on a product that it genuinely came from a responsible
source, no matter how many links in the supply chain
there might be. So the FSC set up a two-part system:
responsible forest management and a chain of custody
(CoC) with labels that would allow the customer to
understand what they were buying. Both elements
are third-party assessed.

The FSC takes a multi-stakeholder approach to forest management and has three equally weighted bodies that debate and agree all decisions: economic, social, and environmental. This allows the voices of business, local communities, and environmental NGOs to be heard equally in all decisions. The FSC acknowledges that different environments and cultural circumstances will have different needs, so while the core standards for forest management, CoC, and labeling are set centrally, a national standard for forest management must be agreed by the FSC group of each country. While each national standard will not be identical, the core values and requirements will be.

The conservation requirements and social responsibility in the FSC standard make FSC the only certification system endorsed by key environmental NGOs: the World Wide Fund for Nature (WWF), Greenpeace, and the Woodland Trust.

The forest management standards of FSC were originally set up to create responsible forestry in the Amazon for large areas of land where the forest has no titled owner, and therefore it was most practical to assume that a country's government is the owner; the land rights of indigenous people were consequently a key issue. This led to a certification structure that initially was not appropriate to countries in which there were many small forest owners. The Program for Endorsement of Forest Certification (PEFC) was created in recognition of the need for responsible forestry and the ability to certify the small forest holdings through a group-

certification approach. PEFC is an umbrella organization and, as an endorsement scheme with central "sustainability benchmarks," is able to reflect the requirements of the country it is operating in, according to its local priorities and forest conditions. Rather than set standards of forest management centrally, these standards are created at the national level and endorsed by PEFC after they have been third-party assessed for compatibility with PEFC's sustainability benchmarks.

There are many similarities between the PEFC and FSC forest management requirements, but while PEFC recognizes the importance of forests, their biodiversity and communities, the origin of PEFC as an organization set up to answer the needs of the forest industry created gaps in its forest management approaches in respect to plantations and land-use change, as well as the rights of indigenous people. However, the two systems have evolved and FSC has group certification available for small forest holders, while PEFC now does not allow certification of plantations where the plantation was converted from a forest after 2010; the land rights of indigenous people are also now addressed.

There continues to be a gap between FSC and PEFC in the chain of custody requirement for a paper to be labeled as certified. Where neither system requires all pulps in a paper to be certified, the noncertified percentage in FSC does need to be Controlled Wood and must therefore go through an assessment process to ensure that the following five origins are avoided:

1. Illegally harvested wood
2. Wood harvested in violation of traditional and civil rights
3. Wood harvested in forests in which High Conservation Values (areas particularly worthy of protection) are threatened through management activities
4. Wood harvested from conversion of natural forests
5. Wood harvested from areas where genetically modified trees are planted

The noncertified portion of a PEFC paper also needs to be noncontroversial, which under PEFC regulation currently equates to being of legal origin. PEFC is reviewing this aspect of the standard and would like to move toward additional sustainability criteria, but at present FSC is a higher standard in this area. The new PEFC standard is due to come into effect in 2013.

Both certification schemes are responding to the due diligence requirement of the EUTR and ensuring that the legality of products labeled through their schemes is clear. FSC is developing an Online Claims Platform that will allow traceability within the supply chain and support companies in their compliance with the EUTR and in reducing risk with regard to the Lacey Act. It is due to roll out in 2014. The comparison of these two certification schemes highlights the role certification has played in promoting responsible forest management, how certification achieves this, and also the important role that certification can play in a due diligence system for buying forest products. There will be further discussion of certification in practice in Chapter 7.

PEFC CERTIFIED

PEFC RECYCLED

PROMOTING SFM

PEFC/16-44-1485
PROMOTING SUSTAINABLE FOREST MANAGEMENT

Right: The direct flow of REDD funds from donor country through to host country. Indirectly, are the host countries still paying for the northern hemisphere's carbon creation?

REDD

The Reducing Emissions from Deforestation and Forest Degradation initiative is a UN program designed to create a value for the carbon storage potential of forests. This initiative has the potential to change or expand the model of responsibly managed forestry.

REDD strategies aim to make forests more valuable standing than they would be cut down, by creating a financial value for the carbon stored in trees. Once this carbon is assessed and quantified, the final phase of REDD involves *developed* countries paying *developing* countries carbon offsets for their standing forests.[11]

There are several potential benefits here:
• Carbon storage and the reduction in carbon emissions
• Forests left intact and the reduction of biodiversity loss
• Creation of a financial flow from developed to developing nations

There are also concerns about the program:
• **Funding:** One proposed option is that it will be funded by offsets. The developed nations purchasing these offsets will not be incentivized to actively reduce their nation's direct carbon emissions. The burden of actual carbon emission reduction continues to be placed on developing nations.
• **Accuracy of carbon accounting in forest:** Inaccurate counting of carbon could lead to a flawed basis on which to sell offsets.
• **Leakage:** Pushing forestry into unprotected areas at both a national and international level.
• **Land rights of indigenous people:** These rights are not included in the valuation of the land; will their exclusion lead to land grabbing for the carbon value?
• **Definition of a forest:** Payment model is based on carbon stored, not on virgin forest left intact. This could lead to plantations replacing forests and continued biodiversity loss.[12]

There is at this point no mechanism that creates forest products that support the REDD process, in that there isn't a mechanism linking the forest areas that will remain intact with well-managed production forests. However, if the concerns about REDD can be addressed there is potential for this approach to be incorporated into the certification schemes with carbon retention becoming part of the environmental values a certification scheme seeks to safeguard.

11. http://www.un-redd.org/AboutUNREDDProgramme/FAQs/tabid/586/Default.aspx
12. http://www.redd-monitor.org/redd-an-introduction/

HOST COUNTRIES

Argentina
Bolivia
Brazil
Burkina Faso
Cambodia
Cameroon
Central African Republic
Chile
Colombia
Costa Rica
Democratic Republic of Congo
El Salvador
Equatorial Guinea
Ethiopia
Gabon
Ghana
Guatemala
Guyana
Honduras
Indonesia
Kenya
Laos PDR
Liberia
Madagascar
Mexico
Mozambique
Nepal
Nicaragua
Panama
Papua New Guinea
Paraguay
Peru
Philippines
Republic of the Congo
Solomon Islands
Suriname
Tanzania
Thailand
Uganda
Vanuatu
Vietnam
Zambia

DONORS

AUSTRALIA
CANADA
DENMARK
FINLAND
FRANCE
GERMANY
JAPAN
NETHERLANDS
NORWAY
SPAIN
SWITZERLAND
UK
USA

FUNDS

UN-REDD
112 million USD

FCPF
250 million USD

FIP
558 million USD

CBFF
150 million USD

UN-REDD United Nations Collaborative program on REDD
FCPF Forest Carbon Partnership Facility
FIP Forest Investment Program
CBFF Congo Basin Forest Fund

Source: Forest, Climate, and Livelihood research network/Westhol, Lisa, Getting ready for REDD+

Water

We are in the early stages of understanding and managing our water use, but what follows are useful methods for understanding the impact of our businesses and products on water availability, and frameworks and tools for managing that impact.

Water scarcity is a highly complex issue that has to be considered on a local basis, as even within a single country there can be some areas that are water rich while others do not have enough to meet the local needs. For this reason footprinting and management at the local level is key to understanding and reducing water scarcity.

The Water Footprinting Network has defined the aspects of a water footprint: the green, blue, and gray footprints. The green water footprint is the rainwater consumed. The blue water footprint is how much ground or surface water is consumed and evaporated in human use: in paper manufacturing that is the water used and not returned to the river it came from. In printing, this would be the water used by the print process. The gray footprint is the pollutants in water returned to its original source.[13] The impacts of the publishing sector can be seen when all four of these elements are taken together.

Water footprinting and water management are comparatively new disciplines, but in the last few years some useful tools for creating a strategy for water management have been created.

Above: Map of water footprint per capita.

13. http://www.waterfootprint.org/?page=files/FAQ_Technical_questions

Water footprint per capita m3 per year (1240 global average)

600-1000

1000-1200

1200-1300

1300-1500

1500-1800

1800-2100

2100-2500

Source: US Infrastructure (www.americainfra.com)/WaterFootprint.org

UN CEO WATER MANDATE

The Water Mandate was launched in 2007 by the United Nations. The mandate recognizes that agriculture has the greatest impact on water availability, but also that business and industry substantially affect water availability and quality, and that these groups can make their impact a more positive one.

The water mandate has six core principles that it asks Chief Executive Officers (or equivalents) to sign up to, endorse, and implement over time:

- **Direct operations:** To assess, set targets, invest in new technologies, raise awareness, and make considerations of water stress a part of all relevant business decisions.
- **Supply chain and watershed management:** To encourage suppliers to conserve and reuse water. To improve the quality of their waste-water treatment. To share best practice with suppliers and to encourage major suppliers to share their targets and reporting.
- **Collective action:** To build relationships with civil society groups and to work with national, regional, and local governments. To encourage the development of new technologies. To support pre-existing water sustainability initiatives.
- **Public policy:** Engaging with other stakeholder groups to contribute to framing policy that will create market mechanisms that encourage water sustainability. Engage in multi-stakeholder groups that address issues of water

sustainability. Promote the role of the private sector in supporting integrated water resource management.
- **Community engagement:** Endeavor to understand local water and sanitation challenges and how operations impact them. Be active in local community and government initiatives that advance the water and sanitation agendas. Support the development of local water infrastructure.
- **Transparency:** To communicate annually all actions and investments undertaken in relation to the CEO water mandate. To publish and share the targets, achievements, and future plans of water strategy. To be transparent in dealings with local government.[14]

This provides a useful framework for shaping a water strategy for direct operations and for a supply chain.

14. http://ceowatermandate.org/files/Ceo_water_mandate.pdf

The Voice of Water

EUROPEAN WATER STEWARDSHIP SYSTEM

The European Water Stewardship (EWS) system is in its infancy, having just launched at the end of 2011, but it has been created with strong input from the EU, from agriculture, and from business groups such as the Confederation of European Paper Industries (CEPI)—and it looks as if it could provide assurances that a business is managing its water use appropriately and to a particular degree.

Water scarcity, as already mentioned, is a local issue, and the EWS looks at water at the river-basin level. The EWS provides a standard, a certification, and a logo. This logo will be applicable to the company's operational sites rather than the products, and there are three logos—bronze, silver, and gold—indicating the level of performance and providing independently verified evidence of sustainable water management.

The EWS system, in addition to providing evidence of sustainable water management to relevant stakeholders, should also provide for the company seeking certification the following:

- Internal benchmark and target-setting device
- Improved and optimized resource management
- A best practice tool
- Steps toward securing future water availability
- Enhanced awareness of the activities and challenges as well as a higher profile within the community around that river basin

- Visibility and access to markets
- A mitigation of physical, regulatory, and reputational risks[15]

Having learned lessons from other certification systems, the EWS will provide a group certification for small-to-medium enterprises (SMEs) around a river basin and also a multi-site communication strategy for the instances in which the multiple sites are located around different river basins.

The EWS system has been developed by the European Water Partnership (EWP), which is an Alliance for Water Stewardship Board organization. The Alliance for Water Stewardship aims to create a globally applicable stewardship system, and as it develops, that program hopes to learn lessons from the EWS. This, along with the input from CEPI in the development of the standard, indicates that this system could have a substantial uptake from the paper manufacturing sector using a widely accepted approach to water management. This makes the system of interest to the publishing sector: it could be used as part of a strategy to establish water management in the supply chain.

15. http://www.ewp.eu/activities/water-stewardship/project-communication-swm

Carbon

Right: Scopes and emissions
in the GHG Protocol.

As the realities of climate change have been acknowledged, significant tools for understanding and reducing carbon footprints have been created, largely in the form of voluntary codes and frameworks, but also as legal requirements.

GHG PROTOCOL

The Green House Gas Protocol is a voluntary reporting scheme and standardization of reporting scopes developed by the NGOs' World Resources Institute (WRI) and World Business Council for Sustainable Development (WBCSD).[16] One of the biggest issues in GHG reporting has been the scope to which a business or organization should report: what activities create carbon, what is a business responsible for, and what can be reported on? The reporting scopes set by the GHG Protocol were created through a multi-stakeholder process including environmental groups (WWF, The Energy Research Institute) and the energy industry (Shell, Tokyo Electric), and the resulting scope has been effective in increasing standardization of reporting as it has been widely adopted. Major businesses report into the scheme, and many others publicly report adopting the scopes of the GHG Protocol. The scope as set out in the GHG Protocol is officially recognized as the basis for carbon accounting and reporting; it is used by many governments to structure carbon reporting and it is the basis for the International Organization for Standardization (ISO) standards that include carbon mapping and reporting.[17]

There are three scopes to report on: scopes 1 and 2 are mandatory (where an organization has chosen to report following the GHG protocol) and are carefully designed to ensure that two companies do not report on the same emissions. The third scope is voluntary.

Scope 1: Direct GHG: Those owned or controlled by the organization: boilers, furnaces, vehicles, and emissions from chemical production.

Scope 2: Electricity indirect GHG: Emissions from the generation of purchased electricity.

Scope 3: Emissions that are a result of the organization's activity, for example, manufacturing and the use and disposal of products.

CASE STUDY: PEARSON

Pearson, one of the world's leading learning companies, reports into the CRC scheme in the UK and publicly reports on:
Scope 1: All fuel used in their buildings and vehicles, including refrigerants

Scope 2: Electricity used in its buildings drawn from grids where it does business

Scope 3: Emissions relating to air and rail travel

16. http://www.ghgprotocol.org
17. ISO, WBSCD and the WRI signed a Memorandum of Understanding in 2007 that they would jointly promote both standards (http://www.ghgprotocol.org/about-ghgp)

CO_2 N_2O CH_4 HFC_s PFC_s SF_s

UPSTREAM ACTIVITIES

1 Purchased goods and services

2 Capital goods

3 Purchased electricity, steam, heating, and cooling for own use

4 Fuel and energy related activities

5 Transportation and distribution

6 Waste generated in operations

7 Business travel

8 Employee commuting

9 Leased assets

REPORTING COMPANY

1 Company facilities

2 Company vehicles

DOWNSTREAM ACTIVITIES

1 Transportation and distribution

2 Processing of sold products

3 Use of sold products

4 End-of-life treatment of sold products

5 Leased assets

6 Franchises

7 Investments

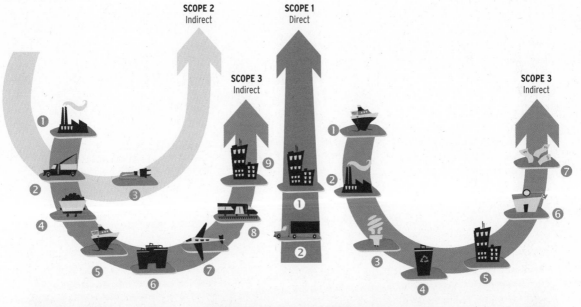

SCOPE 2
Indirect

SCOPE 1
Direct

SCOPE 3
Indirect

SCOPE 3
Indirect

UPSTREAM ACTIVITIES   REPORTING COMPANY   DOWNSTREAM ACTIVITIES

Source: The Greenhouse Gas Protocol

CRC ENERGY EFFICIENCY SCHEME

Previously known as the Carbon Reduction Commitment (CRC), in 2010 this scheme made it mandatory to register and report into the scheme for organizations based in the UK and using over 6,000 megawatts of half-hourly metered electricity per annum. That's an energy bill of about £500,000 ($750,000). As of 2012, the incentive in this scheme to become more energy efficient is that energy allowances must be purchased from the UK government to cover the previous year's emissions, and the less energy used the fewer purchases will need to be made. In addition to this, the organizations registered in this scheme will have their performances rated in a publicly available league table, and improvements here will reflect well and improve reputations.

The scheme is limited to the larger energy users (outside the specifically energy-intensive industries whose emissions are already covered by the 2001 Climate Change Agreements), and therefore targets the areas most in need of reduction. This includes the companies with more substantial infrastructures who will be able to absorb the requirement to produce this reporting and implement plans to reduce their overall carbon emissions.[18]

EPA GHG REPORTING PROGRAM

The U.S. Environmental Protection Agency (EPA) has a mandatory GHG reporting program for large emitters of carbon and other greenhouse gases. This program is for the purpose of informing future policy, and as of 2012 the reporting is available to the public.[19] Two types of emitters are included in this program: "direct emitters" of GHG, and "suppliers" whose products in use emit GHG. The EPA program therefore covers the energy industry. Pulp and paper manufacturers are included as direct emitters and are one of the nine industry groups required to report. These are important frameworks and reporting programs; they set the terms in which we start to understand our energy use and our carbon creation through business activities. However, we are only at the beginning of reducing our carbon creation and these reporting schemes would need to be greatly expanded to start making a substantial impact. Where frameworks for reporting and reduction exist, such as the GHG Protocol, it is worthwhile implementing these systems now to protect the planet, and to future proof your business in preparation for future regulation.

CEPI 10 TOES

CEPI (Confederation of European Paper Industries) has created what it hopes will be a widely adopted framework for accounting for the carbon used to produce paper.[20] It outlines the ten key elements or "Toes" that make up the carbon footprint in the paper making process and gives guidance on how to calculate each element.

As an industry initiative it can be seen as a bottom-up approach, with the industry feeding in its knowledge about the different stages of the manufacturing process

18. More information about the CRC Energy Efficiency Scheme can be found at the Carbon Trust website: http://www.carbontrust.co.uk/policy-legislation/business-public-sector/pages/carbon-reduction-commitment.aspx

19. http://www.epa.gov/climatechange/emissions/ghgdata/index.html
20. http://www.ncasi.org/support/downloads/documents/CEPI_Carbon_Footprint_report_2007.

THE 10 "TOES" OF A CARBON FOOTPRINT FOR PAPER AND BOARD

1. Carbon sequestration in forests

2. Carbon in forest products

3. Emissions from manufacturing

4. Emissions from fibre production

5. Emissions from production of other raw materials

6. Emissions from energy production

7. Emissions from transport

8. Emissions from the use of products

9. Emissions from product end-of-life

10. Avoided emissions

and how best to account for each stage. Using officially recognized frameworks for footprint calculation, such as the GHG Protocol, the 10 Toes produce a framework for the industry that can be accepted by customers and NGOs or for government reporting.

In using this framework the communication of the results would be crucial as the 10 Toes represent the stages of the product's life that affect carbon levels, but the framework does not require that all should be included. Most footprints using the 10 Toes do not include

the first two toes that relate to the effect of forestry on carbon levels since this is an area that has not been fully resolved as to the way it should be included. The GHG Protocol discusses that forest sequestration could be incorporated into a footprint in principle, but no accepted methodology has been developed yet. Because the carbon stored in a tree is not released in the making of paper but is stored in the final book, the carbon footprint could be a negative rather than positive figure; that is, the figure could become minus X tons rather than plus X tons, which would mean that carbon had been taken out of the atmosphere rather than added to it. While it is the case that paper stores carbon, it is not necessarily helpful to produce an overall figure that hides the carbon-producing effect of manufacturing. CEPI included this stage in its 10 Toes to make the issue widely known and acknowledged.

ISO 16759

This International Standard is expected to be released in 2013 and will provide a framework for carbon calculators measuring the emissions from printed media. The standard will be applicable to all sectors involved in the creation of printed media, from prepress to print.

The framework is to be as comprehensive and flexible as possible so as to be globally applicable, covering all print methods, and all finishes, as well as different print-run lengths. This would allow the buyers of products to compare the carbon impacts of a traditionally printed book with a print run of 2,000 to a digitally printed book

"ISO 16759 aims to capture all prepress, print processes, and materials used in the lifecycle of a printed product, be they magazine, pamphlet, or board book."

with only 200 printed where the information had come from the same calculator or calculators built using identical parameters. The standard aims to capture all prepress, print processes, and materials used in the lifecycle of a printed product, be they magazine, pamphlet, or board book. Not only must the processes and materials be captured, their characteristics must be outlined so that the basis for the carbon footprint can be fully defined.

The resulting carbon calculators will be complex tools with many areas that must be considered and included or excluded, to ensure comparability results should come from the same calculator. The standard makes it clear that transparency, and readily available notes accompanying the calculator—on inclusions, exclusions, and normalizations—is required. This standard is an implementation of ISO 14067, which sets out the methodology for building carbon calculators.

In this chapter we have looked at the environmental concerns for the publishing industry and at the initiatives that aim to tackle these problems. We have seen the complex and potentially effective interplay between the private sector, government, and NGOs, all of which have a valuable input in finding solutions to these problems via specialist knowledge of the environmental issues at hand and the level of governance needed to resolve them. All of this sets a background to the publishing industry's response to these environmental concerns, which we will look at in the next chapter, including tools created by the industry and widely adopted standards such as ISO 14001.

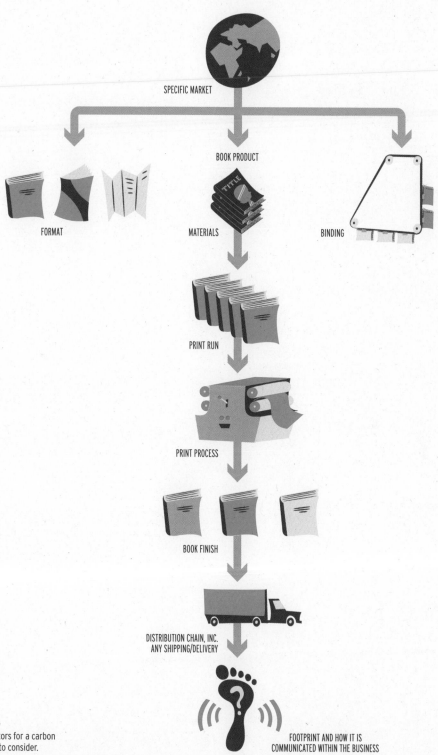

SPECIFIC MARKET

BOOK PRODUCT

FORMAT

MATERIALS

BINDING

PRINT RUN

PRINT PROCESS

BOOK FINISH

DISTRIBUTION CHAIN, INC.
ANY SHIPPING/DELIVERY

FOOTPRINT AND HOW IT IS
COMMUNICATED WITHIN THE BUSINESS

Right: Factors for a carbon
calculator to consider.

The environment &
the publishing sector

We have now examined the major environmental concerns that affect any industry manufacturing and using forest products, and at the initiatives that are trying to tackle these concerns. But each sector with its unique products will have its own specific concerns and answers to them. We will now look at the most pressing environmental issues for the publishing sector, its response to them, and how we can start to use these tools to create a strategy for our own organizations to understand and reduce our environmental impacts. Crucially, we will also look at the business benefits of a strategic response to environmental concerns.

Environmental concerns in the publishing sector

The priority of the publishing sector is a need to source well-managed papers, and the aim of a paper purchasing policy will be to use certified papers that are third-party assessed, ensuring both legality of the pulp fiber origins and that those pulps come from well-managed sources. The most pressing concern for the publishing sector has become finding paper supplies that are not contributing to harmful forestry practices where certified papers are not available. Certified papers have been a great thing for the publishing industry as the work of guaranteeing a level of responsibly managed paper has been done for us, but only a small percentage (around 13%, and mostly in the global north) of the world's production forests are certified, and there is not enough certified-produced paper available to meet the needs of the publishing industry.

How then do you go about finding papers that won't be placing too great a burden on the environment?

The supply chains of the publishing sector are long, which is similar to many industries, but the make-up of paper creates a unique problem, as the formation of the necessary paper characteristics—flexibility, strength, thickness, and so on—requires different varieties of pulp; that is, pulps made from different species of trees, which will be grown in different geographical locations. A softwood fiber may come from Finland and a hardwood fiber may come from Chile. There are enough different paper varieties in the world to require a huge number of paper recipes, and these recipes may require anything from three to thirty pulp sources, each from a different forest. An extensive supply chain.

Above: Many papers require both hardwood and softwood, and pulp is delivered globally to meet demand.

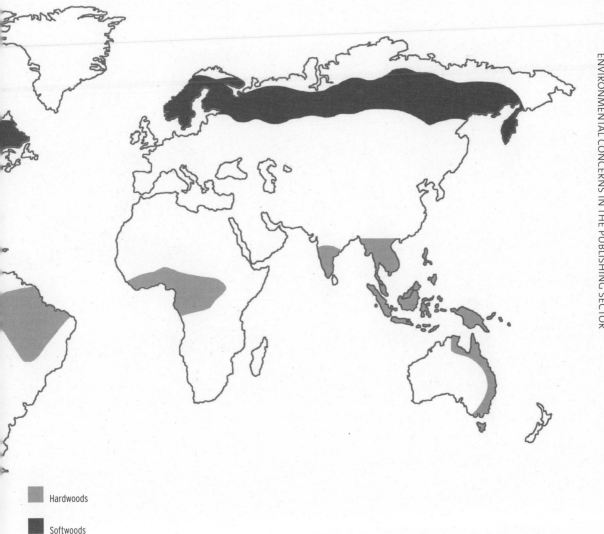

Hardwoods

Softwoods

"The make-up of paper creates
a unique problem, as the
formation of the necessary paper
characteristics requires different
varieties of pulp."

Sourcing low-risk, uncertified papers

PUBLISHER

PREPS

A large quantity of work would be required to ensure, as individual businesses, that the pulps in the papers you use are from low-risk sources, so a group of publishers in the UK has sought a collective response to this problem. They have created PREPS, the Publishers' database for Responsible Environmental Paper Sourcing. The idea behind it is quite a simple one: all the papers used by the publishers are submitted to a secretariat that runs the database. The mills producing those papers are then requested by the secretariat to submit their pulp sources, including the tree species and the forest country of origin. These sources are then risk assessed and graded as high-risk, low-risk, or responsibly managed (responsibly managed requires certification—for more information about the PREPS grading system see Chapter 5). Recording and risk assessing each pulp source can accurately determine whether or not the paper has come from high-risk or well-managed sources, allowing those who access the database to make informed choices about the papers they will use in their publications, even where they are not certified.

In sourcing responsibly managed and low-risk papers, the PREPS database is a highly useful tool. It can also contribute to a due diligence process through its risk assessment of the individual pulp sources. With the increasing requirement of the forest products industry for due diligence from the U.S. Lacey Act and from the EU Timber Regulation, a due-diligence process is becoming vital for all tree-derived materials, and it is of particular importance for those papers that do not have certification and therefore are not bought with a third-party assurance of the legality of the pulps within that paper.

The main focus of the PREPS database is to gather information on the forest sources of pulp, as this is where the chief impact of paper lies, as well as being the least transparent area due to its complexity. But, acknowledging the multiple impacts of any manufacturing process, the database is now starting to incorporate information on the carbon impacts and water use of the mills that it gathers information from.

To ensure that the information PREPS provides here is both relevant and in line with data gathering from other industry carbon footprints, the carbon indicators in the CEPI 10 Toes, Paper Profiles, and International Council of Forest and Paper Association frameworks have all been referenced. Using these already established frameworks should also mean that no work on the supplying mills' part need be duplicated or re-worked.

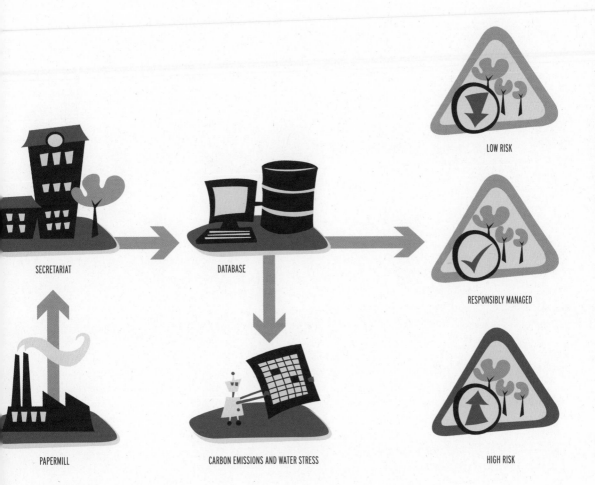

SECRETARIAT

DATABASE

LOW RISK

RESPONSIBLY MANAGED

PAPERMILL

CARBON EMISSIONS AND WATER STRESS

HIGH RISK

Above: The flow of information through PREPS.

These are not full lifecycle footprints, but they do allow for a greater understanding of the mills' environmental impacts and can be used by publishers to inform their purchasing decisions.

The water information gathered will also be based on existing tools and footprint methodologies from Paper Profile, WWF Paper Score Card (now Check Your Paper), EPAT (Environmental Paper Assessment Tool), the Water Footprinting Network, and the WBCSD Global Water Tool, and again it will provide information for comparison rather than the full footprint.

Using pre-existing methodologies reduces duplication of work in the burgeoning green reporting movement. It is important that all these methodologies are transparent and communicated, so that it can be understood what the final figures represent, creating clarity for the end users.

Environmental impact mapping

The inclusion of carbon and water information by PREPS incorporates the second key concern for the publishing sector, which is to map the impacts of publishing by using lessons learned from the global initiatives and tools aimed at understanding and reducing the effects of human activity on the environment. Publishers are mapping their impacts and producing—using widely accepted methodologies—tools that will enable ongoing reporting and build as full a picture as possible of those impacts for individual products and for a business. By using widely recognized standards to base reporting on, the publishing sector is preparing itself for the possibility of increased mandatory environmental reporting in the future. We will now look at the tools and framework standards the publishing industry uses to understand, communicate, and reduce its impacts.

CARBON ACCOUNTING

As mentioned in Chapter 1, there are several globally recognized standards that are designed to show organizations how to account for their GHG emissions, including carbon. The UK Environmental Action Group (EAG) of the Publishers Association and Booksellers Association has produced two carbon accounting tools. The first is an online tool that allows all its members to translate their direct impacts as set out in the GHG protocol and convert these to GHG emissions. This tool will aid small-to-medium enterprises (SMEs) in reporting on, accounting for, and offsetting their carbon emissions.

It is worth mentioning here that the goal is to reduce carbon emissions as much as possible, and that offsetting is useful for emissions that cannot be lowered further and, in the short term, as a financial driver to lower emissions. Offsetting cannot be relied on long term as it does not involve an overall lowering of global emissions.

The second tool to come out of the EAG is the Bookcarbon Calculator. This calculator looks at the carbon emissions for the lifecycle of a book from manufacture of the pulp and paper to delivery of the book to the customer-gate. This carbon calculator accounts for publishers' indirect emissions through the books they have made. The calculator allows users to vary materials, the format, the printer, the print method, and the print-run length; the results can be used for reporting purposes in preparation for expanding regulation in this area. This also allows users to see the carbon effects of particular print decisions, making visible opportunities for carbon reductions.

The tools outlined above have been created by the UK publishing sector for the UK publishing sector, but they are a good example of what can be done collectively by pooling resources (time and funds) and harnessing the knowledge held across a group. Using a shared calculator creates comparable reporting, which increases the transparency of the industry.

If you are involved in print production but not directly for the publishing industry, your own industry representative body may have similar tools, and it is worth making enquiries.

"Using a shared calculator creates comparable reporting, which increases the transparency of the industry."

WWF'S CHECK YOUR PAPER

WWF has created the Check Your Paper website, a database in which you choose your paper type and can then review the information held on specific paper products. The tool assesses the key environmental impacts and each paper is rated in six areas:

• Fiber source
• Fossil CO_2 emissions from manufacturing
• Waste to landfill
• Water pollution from bleaching
• Organic water pollution
• Environmental Management System

The aim of this system is to promote responsible paper procurement by making the choices available clear, with an emphasis, via the rating system, on certified and recycled fiber sourcing, because it is in the fiber sources and the potential deforestation risks where the most substantial impact of paper production lies. The fibers must come from one of five sources:

• Recycled post-consumer waste
• Recycled pre-consumer waste
• Virgin fiber of legal origin
• Virgin fiber from controlled sources
• Virgin fiber from credibly certified sources

Each paper in the database is third-party audited, and the verification of the legality of the fiber sources will come from a credible certification system (FSC and PEFC) and national and international legality schemes, with wood covered by a FLEGT VPA (see Chapter 1 for a discussion of certification systems and the FLEGT VPA). The database encourages transparency in the supply chain and demonstrates the work that paper manufacturers are putting in to reduce the negative environmental impacts of the paper manufacturing process.

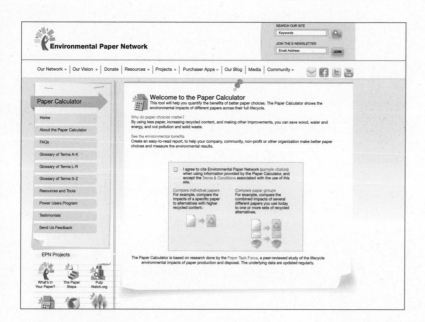

Left: The Environmental Paper Network's Paper Calculator.

A LIFECYCLE APPROACH

Two significant initiatives in the U.S. take a lifecycle approach and look at the impacts of paper production and disposal, and book manufacture and disposal, respectively.

The Paper Calculator was created in 2005 by the Environmental Defense Fund alongside the Paper Task Force (a voluntary private-sector initiative): it uses a database of information collected by the industry. The calculator was third-party reviewed by scientists and environmental NGOs in 2008; from this consultation additional information was added in 2009. Given information about the type and quantity of paper used, the calculator can show you how much wood was used, how many trees this equates to, how much water and energy was used, the weight of VOCs (volatile organic compounds) that were used, and other factors as well. These figures are based on the national production averages in the U.S. The tool is web-based and can be used by anyone to show the impact of the paper they have used. It can also be used to "design-for-the-environment" as the impacts become visible (more on this idea in Chapter 4).

The Book Industry Environmental Council (BIEC) is a representative network of the book publishing industry equally representing publishers, book manufacturers, paper manufacturers and suppliers, booksellers, and environmental NGOs to establish best practice and communicate it to the industry. BIEC's goal is to benchmark, track, and improve the environmental footprint of the book industry, and it is doing this through three main areas of focus: tracking the impacts of the industry; setting goals for the reduction of carbon emissions; and reducing returns. BIEC has surveyed its members and uses this to benchmark and track the publishing sector's environmental impact and is in the process of developing a strategy for reducing returns and increasing book recovering and recycling (we'll look at this topic in Chapter 9). For the reduction in carbon emissions, BIEC has produced a strategy on how to achieve this goal, as well as guidance for publishers, printers, and paper makers on how to put that strategy into practice. These guides are available on the BIEC website (see the resources appendix for this address).

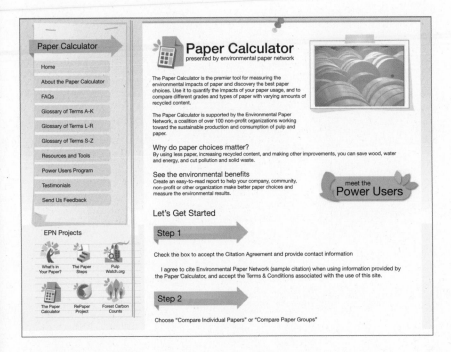

Left: Compare papers
or paper groups.

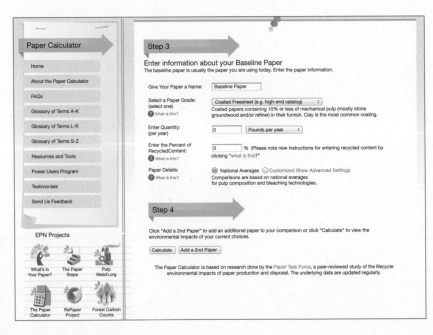

Left: Use the calculator
to understand the effects
of your paper purchasing.

The ISO 14001 process

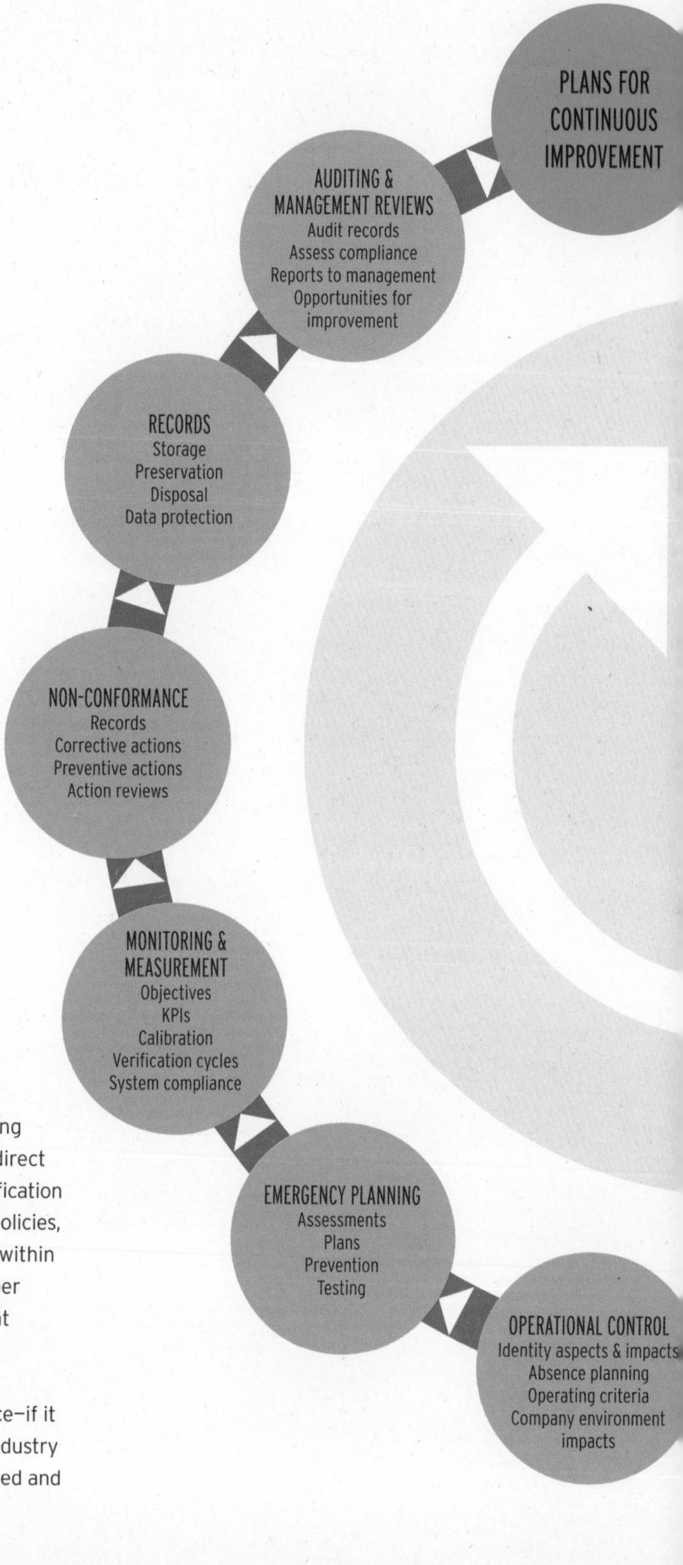

PLANS FOR CONTINUOUS IMPROVEMENT

AUDITING & MANAGEMENT REVIEWS
Audit records
Assess compliance
Reports to management
Opportunities for improvement

RECORDS
Storage
Preservation
Disposal
Data protection

NON-CONFORMANCE
Records
Corrective actions
Preventive actions
Action reviews

MONITORING & MEASUREMENT
Objectives
KPIs
Calibration
Verification cycles
System compliance

EMERGENCY PLANNING
Assessments
Plans
Prevention
Testing

OPERATIONAL CONTROL
Identity aspects & impacts
Absence planning
Operating criteria
Company environment impacts

ISO 14001

ISO 14001 certifies that an organization is addressing all the environmental impacts within a company's direct control and sphere of influence, over time. This certification is an indication of an organization's environmental policies, and there is a high level of uptake of this standard within the publishing supply chain: many printers and paper manufacturers have an Environmental Management System (EMS) that has been certified as being ISO 14001 compliant.

The standard does not set levels of performance—if it did a separate standard would be needed for each industry group—but requires that all impacts should be mapped and

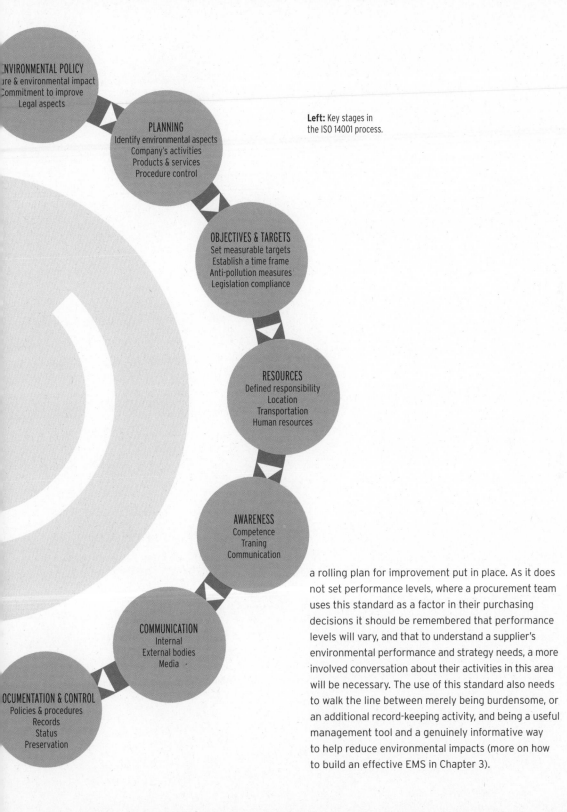

ENVIRONMENTAL POLICY
ure & environmental impact
Commitment to improve
Legal aspects

PLANNING
Identify environmental aspects
Company's activities
Products & services
Procedure control

Left: Key stages in
the ISO 14001 process.

OBJECTIVES & TARGETS
Set measurable targets
Establish a time frame
Anti-pollution measures
Legislation compliance

RESOURCES
Defined responsibility
Location
Transportation
Human resources

AWARENESS
Competence
Traning
Communication

COMMUNICATION
Internal
External bodies
Media

OCUMENTATION & CONTROL
Policies & procedures
Records
Status
Preservation

a rolling plan for improvement put in place. As it does
not set performance levels, where a procurement team
uses this standard as a factor in their purchasing
decisions it should be remembered that performance
levels will vary, and that to understand a supplier's
environmental performance and strategy needs, a more
involved conversation about their activities in this area
will be necessary. The use of this standard also needs
to walk the line between merely being burdensome, or
an additional record-keeping activity, and being a useful
management tool and a genuinely informative way
to help reduce environmental impacts (more on how
to build an effective EMS in Chapter 3).

Benefits for business

There are some important benefits to your business in addressing supply chain (indirect) impacts: it gives visibility of additional regulation and changes within the marketplace, it strengthens supply chain relationships, and it informs potential product development. Addressing these issues will give you the information you need to develop a due diligence strategy and comply with current regulation as well as prepare you for future regulation. Reducing your impacts through resource efficiency and responsibly managed sources of material will help you to understand the pressures on natural resources and the effects on the ecosystems they come from, allowing you to prepare for a less stable marketplace through your supplier relationships. To address these environmental issues you will work closely with your suppliers, who will be important sources of information for you to be able to map and calculate your organization's environmental impacts; they will also be sources of expertise when developing new products for maximum green potential.

Just as an organization wishes to reduce its environmental impacts, its stakeholders will have the same concerns. There is now great interest in the environmental performance of organizations coming from different quarters, from shareholders and financial markets (as expressed by the FTSE4Good index and the

The development of green technology is marked by drivers and benefits. Among the drivers that are impacting upon major industrial processes, such as printing, are:

1. Tighter regulation on all businesses.

2. The development of green technologies, which enable companies to meet the regulations and improve their carbon footprints.

3. The demands by customers and society as a whole for more eco-friendly industry.

When industry takes on board these drivers, and implements them within their Environmental Management Systems, they create many benefits for themselves, their employees, and society as a whole:

4. They considerably reduce their waste and day-to-day expenditure through efficiency savings in the use of resources and energy.

5. They improve working conditions and reduce the impact of their activity upon the environment by replacing harmful ingredients in the raw materials of printing.

6. They enjoy the cost savings of greater economies of scale as a result of a higher uptake of eco-friendly raw materials, such as paper.

7. They have a competitive advantage as a green manufacturer over non-green companies.

8. They reduce and eliminate waste and carbon dioxide from their processes being released into the atmosphere.

Right: The publishing value chain.

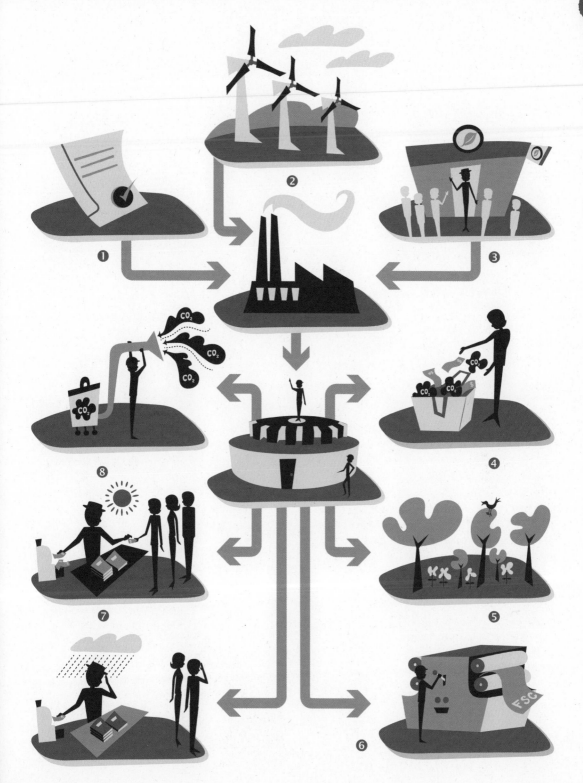

green performance indicators used by investment groups) to governments with increasing regulation requiring reporting and due diligence. Also, NGOs, representing the interests of the people and habitats in the areas resources are taken from, wish to know about an organization's environmental performance and policies and in turn communicate this information to potential customers. Fully addressing the environmental impacts for your organization will allow you to respond to and engage fully with all these stakeholders, benefiting your reputation with potential investors, and creating a relationship with NGOs in which you learn from each other and create due diligence approaches that are effective in practice and reach a high level of governance.

The tools and initiatives in this chapter have been created using industry knowledge, informed by widely accepted methodologies and often with input from NGOs, to respond to the concerns of the publishing sector. The approaches they use can become part of a strategy to source papers that use pulp fibers from responsibly managed forests, ensuring that your products are not supporting deforestation, and also satisfying the due diligence requirement for papers made from legal sources for products delivered into the U.S. and in the future, into the EU. In addition, there are initiatives that aim to make the carbon and broader environmental impacts of the print production process visible; the information learned from these can be used to guide where and how these impacts can be reduced.

We have seen that the main concern of the publishing sector is to make their supply chain and its effects visible and to use this information to source materials responsibly and reduce the industry's impacts. In the next chapter we will look at how to begin making a business greener and creating a management system to help do this.

"Reducing your impacts through resource efficiency and responsibly managed sources of material will help you to understand the pressures on natural resources and the effects on the ecosystems they come from, allowing you to prepare for a less stable marketplace through your supplier relationships …"

Bottom up & top down

The eco-road has many starting points. Discover what actions you can take as an individual to inspire change, whether you are an intern or a business leader. Find out how to structure environmental management systems to develop a consolidated and lasting approach, formalizing processes and continually improving performance. Use green checklists to assess your own company and to calculate the environmental footprints of your business and products. This chapter is about finding ways to put your business on a more ecologically sound footing. We'll examine how change can come about and how you can inspire that change, whatever your position in the organization. We'll talk about how to build a lasting and effective environmental management system (EMS), and provide a checklist to begin the assessment of your business.

Green champions network

How does an organization begin to green itself and how can you as an individual help that process? There is no one route to a greener company, but we do know that changing an organization and making it greener requires collective action and a broad buy-in from its members; it requires a change in the business culture of that organization. A green champions group can initiate this change.

CREATING A GREEN CHAMPIONS NETWORK

Our places of work are communities; we spend around eight hours a day there, and just as we use resources and create waste at home so we do at work, too. And for as many people who care about how they use resources at home, they care about this at work as well. A green champions group is a chance for the people who want to help, to get involved, and to make a difference to their environment at many levels: in their work space, in the local community, and for the wider world. The group also provides a way to initiate change in your organization at a grassroots level.

HOW A CHAMPIONS NETWORK CHANGES A BUSINESS CULTURE

A champions network provides a large base from which the green goals of a business can be directly communicated. These goals can be disseminated through the network members' actions; for example, they communicate tacitly, by enabling recycling through the provision of the correct facilities, or directly, through advocating green policies and having discussions about those policies within their own departments, for instance at team meetings or staff inductions. After all, an idea from a friend is much more likely to be taken up than a general directive received by email. Through this communication the culture of a business is gradually changed. This network can also provide a flexible forum of ideas, where the mass of the members understand that they can attend when they have the opportunity, other work allowing, and where they can share and explore their ideas and know that the network will also have the capacity through time and manpower to implement the ideas. The network can be seen as having elements of a top-down approach, in that central goals are communicated by the group, or a bottom-up approach, in that ideas can come from anywhere in the business, and a horizontal flow of ideas is received from the departments in the form of feedback.

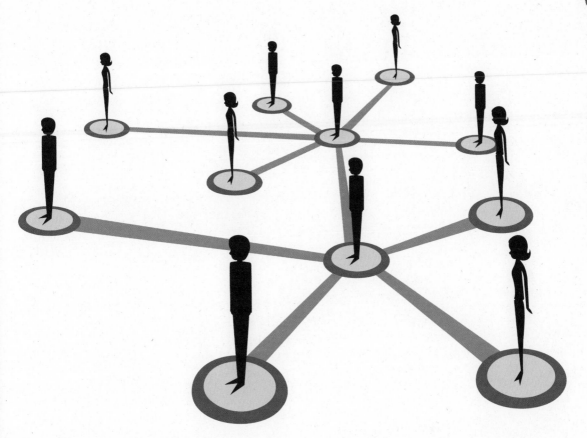

STRUCTURING THE NETWORK

Anyone in an organization could initiate this group. Here are some pointers for the structure of the group that can aid its effectiveness:

- Members can be from any level of the organization: It should be about who is driven and wants to help, but can also include members who are nominated, as motivation can come from wanting to help the cause and from wanting to be more involved in the business.

- One member from each department: This will give you a network for direct communication across the organization, and it will also increase the acceptance of the goals and perceived legitimacy of the network as each department is aware that it is represented.

- A core group of specialists: This core group will attend all the time and help with knowledge of who in a business to approach about a project, as well as provide any necessary information such as spend and usage figures.

- A member of senior management who can communicate and promote the group's goals to the entire senior management level: This can help pave the way to wider company communications, with all staff members seeing that this policy direction has management backing.

- Think about who in your business you will want to share your ideas with: Facilities, operations, buyers? These roles could be involved in the network or the network could support some of their functions.

- A revolving chairperson: Someone to focus the group is useful, but a revolving chair will create an atmosphere of greater inclusivity.

Above: All levels of the organization can be involved in the green champions network.

CASE STUDY: REED ELSEVIER

Reed Elsevier is an education publisher providing learning solutions for professional customers. The Reed Elsevier Environmental Standards program was created as a way to circulate environmental results at a site level, in order to compare the performance at each site with standard levels and to inspire competition between locations. The Standards program covers all global sites that report primary environmental data (118 in 2011) and covers ten areas: Reporting, Certification, Travel, Energy, Climate Change, Water, Waste, Office Paper, Data Center Performance, and Awards. There is also an additional standard awarded, Additional Recognition, for sites that go beyond the standards set and take on further projects, for example, sites setting emissions targets for their car fleets or switching to renewable electricity.

The Standards results are calculated from the previous year's environmental data, and the results are officially released to all employees on World Environment Day (June 5), when the CFO writes an email to all employees about the environmental program and performance. A certificate (signed by the CFO) is awarded to each location that achieves "Green Status" by achieving five or more of the Standards.

An environmental task force helps identify and support any site that is struggling to achieve the standards. This task force collects and distributes best practice and develops processes that can be applicable across many sites, enabling all sites to reduce their environmental impacts and meet the targets set.

IMPLEMENTATION: GOALS AND STANDARDS

The group will have ultimate goals on impacts they want to reduce and how best to reduce those impacts. A project will require a baseline—for example gallons of water used per annum or gallons of water used per staff member per annum in 2013—and a goal that can be communicated. A good way to reach that goal is to set a standard. A standard, as opposed to a target, will say not that this where we plan on getting to but that this is what is achievable, and we are all expected to get there.

If there is a feeling in your organization that setting standards across all operating companies doesn't fit with an autonomous ethos, how about a forum for the sharing of ideas and practices based on a framework that targets key areas of environmental impact? Depending on the structure of your organization this may present more or less of a challenge.

If your business is based over multiple sites, and also across several countries, the setting of standards may present more of a challenge in some areas than others. In this instance the time frame over which you set objectives can be made more flexible, but maintaining that ultimate aim and understanding that setbacks always happen will be vital. But this is where a feedback mechanism may come in useful as information coming back from departments may lead to new ways to implement a project or just an expansion of the method you are already adopting. You are lowering environmental impacts, and this requires a

STANDARD GALLONS OF WATER USED BY A SINGLE MEMBER OF STAFF = APPROX 150 GALLONS PER YEAR

GOAL GALLONS OF WATER USED BY A SINGLE MEMBER OF STAFF = APPROX 130 GALLONS PER YEAR

 = 10 GALLONS

CASE STUDY: COMMUNITY–PENGUIN

Penguin is a globally recognized trade publisher, publishing everything from classics to cookbooks. With the support of the company, Penguin employees give time and money to their local communities and to good causes. Over 1000 Penguin employees around the world take part in the global Penguin Walk raising money for environmental and social causes. Employees volunteer their time to help struggling readers in schools local to Penguin's UK offices.

culture change. And, as not all business cultures are the same, even on a site-to-site basis within an organization, it will take varying amounts of time. That this change should be embedded throughout the business is what the green group is there to help with.

BENEFITS OF REDUCING DIRECT IMPACTS

1. Morale: Through this network you can begin to work on your direct impacts that are easily controlled. Finding projects that can be easily or quickly implemented can be an enormous morale boost for the group and the organization, as there is a genuine sense in which it pleases people to do the right thing for the environment.

2. Cost reduction: The network aims to change the culture of the organization, to reduce impacts through well-managed resources, and for this to become the norm. And there are frequently financial benefits to a reduction in impacts: reducing the amount of energy and fuel used by a business will reduce costs at the same time.

3. Staff retention: Acknowledging further the communal aspect of work through the champions network can increase staff satisfaction and retention. A network that allows and encourages all its members to contribute ideas and actions that benefit everyone empowers these people and creates an atmosphere of inclusivity. The atmosphere of inclusivity is expanded even for those not directly involved in the network through knowing that a person's position in the company in no way prohibits their ability to be involved in shaping and creating change in that company. Increasing staff retention reduces the costs of staff recruitment and training.

4. Community: There are opportunities for building on the sense of community created by the champions network and working on projects with, and in support of, the local community and its environment. Often such projects foster and reinforce a sense of value for the environment for the people who get involved in them.

5. Create knowledge: The process of assessing and managing impacts can help a business understand its indirect impacts through its products and the kind of process that its suppliers will be going through. This knowledge will be beneficial when creating a management system for those altogether less visible indirect impacts.

6. Sustainability: By implementing green goals through a champions network and changing your business culture, you are adding social and environmental value to your business. These values add economic value: as you address the needs of your staff you boost morale, increasing staff retention, and as you reduce your energy use you reduce costs and stabilize your business against fluctuation in the energy markets. Organizations will be financially successful when they respond to environmental and social needs.

Reducing your office impacts

Below: Simple energy saving solutions can be highly effective.

We have looked at how an organization's business culture can be changed to embed green thinking into its decision-making process, creating environmental and social capital. We will now address five areas in which you can start to reduce those impacts, highlighting how they will benefit your organization economically, followed by a simple checklist bringing together all the action points.

ENERGY

Each way in which you reduce your energy use will save your organization money from its energy bill. Some methods of reducing your energy consumption are simple and would be well implemented by a green network, such as switching out lights in rooms when not in use, switching off office equipment when not in use overnight, and lowering the temperature on the thermostat by a degree or two.

The green network can train staff in green office practices. The savings from such actions, though small, will accumulate to something bigger in terms of cultural change and capital. The goodwill and savings generated could be used to fund these longer term projects: installing motion sensors on lights; installing timers to switch off office equipment when not in use; installing energy monitors to show the energy use and cost of equipment; replacing older office equipment with more energy efficient models; and rationalizing data storage.

IT equipment is a large source of energy use in an office, and old office equipment could be replaced with more energy-efficient models. Office equipment such

CASE STUDY: SIMON AND SCHUSTER ON ENERGY SAVING

Simon and Schuster is an international trade publisher of fiction and non-fiction with divisions in the U.S., Canada, Australia, and the UK. Across its divisions, Simon and Schuster employs many power saving techniques.

In the warehouses:
- motion sensors on lights ensure that only areas being actively used are lit
- modern forklift trucks use the momentum of braking to charge their batteries

In the offices:
- desktop and laptop computers are shut down at night
- power save mode is set to switch off monitors and hard drives, and screen savers are set to "none"
- printers and photocopiers are turned off at night
- CRT screens are being phased out and replaced with LCD flat screens

Simon and Schuster Australia estimates a 15% power saving since initiating such measures.
Source: http://www.simonandschuster.biz/corporate/eco-friendly-practices

SWITCH OFF SAVE ENERGY

Left: Monitoring technology allows you to use energy as and when it's needed.

as computers, printers, and servers all have an expected lifespan and, after a certain amount of use, wear, and tear become less functional. Because of this, there is often a replacement cycle for office equipment. You may wish to replace your equipment with more efficient models at the natural point in this replacement cycle, or the energy readings on the equipment may tell you that it would be more cost effective to replace it sooner, as any spend brought forward or increased would be recouped in energy savings.

The energy used by servers for data storage is substantial, and any efficiency made here is a cost saving. You can rationalize the amount of data you store—fewer bytes means less energy and more savings. Holding redundant data on servers has a larger impact than most

users would imagine; not only is it taking up space on a server and leading to increased storage requirements (extra spinning hard disks), but this excess data is included in a backup dataset that can double or treble the storage requirements as companies usually have one backup onsite (duplicate disks) and at least one backup data set offsite (again duplicate disks, tapes, or other removable media). This saving will only be realized fully if you can make part of a server redundant. You may find it appropriate for your organization to move to cloud computing. Using cloud computing, your data would be stored by a server system that serves many customers. Instead of estimating how much server space will be required, you occupy server space as and when needed, including backups, and access that data remotely via the internet.

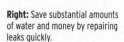

Right: Save substantial amounts of water and money by repairing leaks quickly.

WATER

Savings in water consumption can be made with the help of the site facilities team. You can communicate why and how your organization will be conserving water through your green networks with strategies like installing a water meter, stopping unnecessary water use, locating and repairing water leaks, installing water-conserving plumbing fixtures, and recycling water.

Installing a water meter will allow you to assess and monitor your water consumption. If you also pay for water on the basis of meter readings, you will consequently benefit from any reductions in water use that you make.

WASTE

There is a twofold approach here: first, minimizing resource waste to save money and reduce the resulting waste; second, ensuring that what waste is produced is repurposed suitably, because what is waste in one situation is a valuable resource in another.

Printers can be set to two-sided but there will also be changes needed to general working practices, with more editing on-screen, and so on. The green network can disseminate the word about why there needs to be less printing, and the network members can also ask for feedback from the teams about additional ways in which to reduce paper consumption. As all departments will have different roles and workflows, there will be different opportunities for reducing waste: for instance, a procurement team may want to stop keeping hardcopies of purchase orders where an information system now holds this data automatically. Could your dependence on handouts and printed emails be reduced if there is access to the office information systems in meeting rooms?

Above: Use telecommunications facilities rather than traveling to meetings.

CASE STUDY: DK

DK publishes highly illustrated reference books for adults and children. In 2011, DK reviewed their data storage and initiated programs to maximize the potential of the storage they had available. Images were converted to different file formats that used less data space but kept the necessary clarity of image. Book files were analyzed, and any data that was extraneous to the print requirements was removed.

The green network can support your staff to increase your recycling percentage with information on what can be recycled and where. The key will be to make the system as accessible and simple for staff members to use as possible.

Additionally, are there ways in which you could divert traditionally non-recyclable waste from landfill? Find out if there is a scheme for converting waste to energy in your area. Could you be sending your waste to be anaerobically digested and turned into heat and power rather than being sent to landfill?

Above and Right: Encourage low impact transport, such as bicycles, and cars that run on used vegetable oil.

TRANSPORT

Becoming greener in this area of your business produces direct savings on travel and fuel, improves the health of your staff, reduces your dependence as a business on oil, and protects you from unnecessary exposure to fluctuations in the fuel markets. The efforts to change attitudes to travel or current business systems will be well rewarded.

The green network can help the business to take on new communications technology by acting as super-users and talking staff through any queries they might have about its operation; demystifying a new technology will go a long way to increasing its uptake.

Adoption of public transport can be encouraged with ticket loans, a car pool scheme could be set up by the organization, and cycling and walking to work will be encouraged if there are adequate shower and bike storage facilities.

By sourcing locally produced biodiesel, often made from used cooking oils, you are avoiding biodiesel that may have been produced as part of a destructive forest clearance cycle, if it is made from palm oil, and you are using a product that would otherwise have gone to a landfill. It is also a cheaper alternative to diesel. If you use a transport agent, it is likely that deliveries are already consolidated, but it is worth noting that all consolidation reduces fuel consumption and costs.

PROCUREMENT

The choices that your organization makes through procurement can have a positive effect on your environmental impacts. These are choices with lowered impacts: using recycled stationery and recycled or certified paper. Much of the paper recycling from an office is high-quality white waste and can be made back into papers suitable for office use. Buying recycled papers creates a market for recycled papers and also ensures that paper that has been recycled then fulfills its function and ceases to be waste.

Though some of these choices may appear more expensive at first, the overall cost may actually be lowered. For example, if the upfront cost of sustainably designed furniture is higher, consider its quality and how long it will last; if the quality is good enough, then the long-term cost is lowered, as it will be longer before the item will need to be replaced.

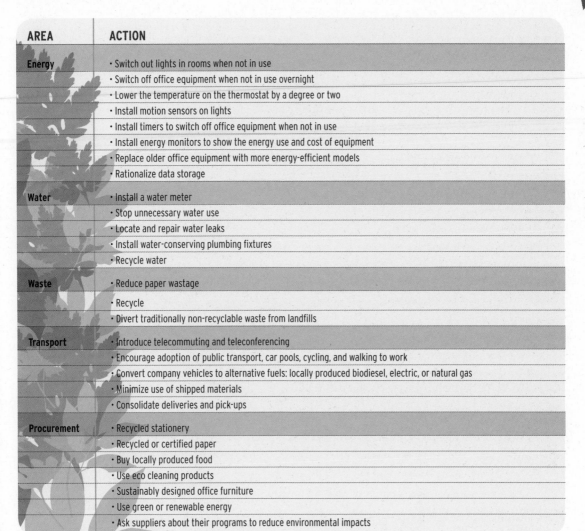

AREA	ACTION
Energy	• Switch out lights in rooms when not in use
	• Switch off office equipment when not in use overnight
	• Lower the temperature on the thermostat by a degree or two
	• Install motion sensors on lights
	• Install timers to switch off office equipment when not in use
	• Install energy monitors to show the energy use and cost of equipment
	• Replace older office equipment with more energy-efficient models
	• Rationalize data storage
Water	• Install a water meter
	• Stop unnecessary water use
	• Locate and repair water leaks
	• Install water-conserving plumbing fixtures
	• Recycle water
Waste	• Reduce paper wastage
	• Recycle
	• Divert traditionally non-recyclable waste from landfills
Transport	• Introduce telecommuting and teleconferencing
	• Encourage adoption of public transport, car pools, cycling, and walking to work
	• Convert company vehicles to alternative fuels: locally produced biodiesel, electric, or natural gas
	• Minimize use of shipped materials
	• Consolidate deliveries and pick-ups
Procurement	• Recycled stationery
	• Recycled or certified paper
	• Buy locally produced food
	• Use eco cleaning products
	• Sustainably designed office furniture
	• Use green or renewable energy
	• Ask suppliers about their programs to reduce environmental impacts

You are on a path to lowering your energy usage, but some energy requirements will remain, and renewable energy and green energy with lower carbon emissions, such as solar energy, are a good option for those energy requirements.

And just as you are on a path to becoming a low impact business and, in the process, learning about where your impacts lie and how to lower them, the same may be true of your suppliers. For example, there has been a lot of work in the data storage industry to increase the use of renewable energy, and also to find new technological solutions to reduce the energy that servers require to make their discs spin and to keep cool. Speak to your suppliers about their products and business, and find out how they are reducing their impacts.

The majority of these projects will save you money by reducing your use of resources. The money saved from the short-term projects can be used for the larger long-term projects that may require a level of investment. All these projects change the culture of an organization, ultimately save money, and stabilize the organization in preparation for the near future in which resources such as energy, water, and wood have become comparatively scarce and more expensive.

All the action points are gathered together in the table above as a checklist for easy reference.

Creating an EMS

Having identified the projects you want to tackle for your organization and some of the methods for putting them into practice via your green network and other specialists, you will want to consolidate that implementation into a company-wide environmental management system (EMS). An EMS is a way of structuring your plans for reducing your impacts that will make record-keeping and reporting easier and keep you on track with your plans. Once an EMS is created, it can be used for managing all an organization's impacts: the direct impacts discussed in this chapter and the indirect impacts created through the supply chain by manufactured products.

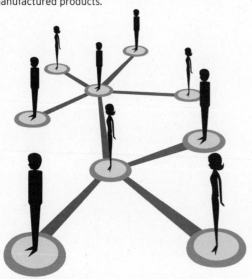

HOW TO STRUCTURE AN EMS

There is one key goal to an EMS:

1. To reduce environmental impacts

But three additional goals will increase the use value of the EMS to your business hugely:

2. To collect the relevant information for reports and communications

3. To reduce financial impacts and increase business stability (EMS as risk mitigation)

4. To record environmental, social, and economic value accrued through sustainable business practices

With these goals in mind, you can create an EMS with long-lasting value to your business.

You can use the checklist as a starting point for an EMS. You will want to include the following elements:

- Area of impact
- Proposed action
- A baseline for each impact
- Targets (communicated as a standard)
- Timeframe
- Who will be involved and responsible for the project
- Balance of costs (expenditure and predicted savings)
- Environmental and social value accrued

Opposite is an example of what it could look like:

"An EMS is a way of structuring your plans for reducing your impacts."

AREA	ACTION	BASELINE FOR EACH IMPACT	TARGET	TIMEFRAME	RESPONSIBILITY	BALANCE OF COSTS	VALUE
Energy	Ask staff to switch off lights in unused areas	$160,000 1200 mega-watts in 2012	10% reduction in electricity in 2012	1 year	Green network	Expenditure $0 Saving $16,000	Boost staff morale if we can communicate quickly the carbon saving
Waste Water	Repair leaks quicker	390,000 gallons in 2012	10% reduction	1 year	Facilities	Expenditure $0 Saving $200	Measuring the improvement and benefit will encourage ongoing action
Waste	Stream waste for recycling and composting	50 tonnes in 2012	70% reduction	1 year	Green network and facilities	Expenditure $300 Saving $130 (expenditure more than recouped after 3 years)	Goodwill created by involving all staff and providing everyone with a green role

Above: Examples of EMS projects and the information needed to get projects off the ground.

Right: Savings made from short-term projects can be used to fund long-term projects.

Below Right: Examples of short- and long-term projects. Long-term projects tend to be those that require a financial outlay or a project that incorporates actively many areas of the business, such as looking at work flows with a view to reducing paper usage.

BUSINESS PROPOSAL

You have now gathered together the information you will need for a business proposal for each project. A business proposal for any project should include: the goal (reduce energy consumption); the baseline of information (for example, energy bill at present); proposed method of reducing impact (campaign to get lights switched off or request for energy sensors on lights so that it happens automatically); cost of project; and the reduction to energy bills and offsetting costs.

ONGOING IMPROVEMENT

The actions on the checklist combine short-term and long-term projects. Many of the short-term projects can be easily implemented, such as turning down the thermostat, and will produce an instant saving on your energy bill; the goodwill and revenue generated from these types of projects can be used to fund and implement the long-term projects. Long-term projects will also yield cost savings but may require an initial outlay of capital, or additional research. Plan your EMS to acknowledge these different types of projects, and you will naturally be led into ongoing improvement as your organization learns about the ways in which it can become greener.

One of the reasons that improvements need to be made over time is that a vital part of becoming a low-impact business is to have the support of the staff all the way through your organization from senior management to the interns. Time must be given to communicate what you want to achieve, why you want to do it, and how everyone can help, and it is rare for any new idea to need communicating just once—there must be a process of reinforcement. This does take time, but it can also create a sense of momentum about the task your business has set itself. As specific ideas become embedded, such as recycling or energy saving, so does the idea of becoming a greener business and ensuring that all resources are used well. In this way, your task of becoming greener picks up pace.

Perhaps one thing to watch out for is that as ideas become embedded and part of the everyday it can start to feel that you are doing less; in actual fact the good work continues and gathering the information on water, energy, waste, and all those other areas can show this.

REPORTING

Whether it's formal business-to-business reporting or communications to customers, for many organizations the EMS will be a source of information for reporting and communicating with internal and external stakeholders on their environmental targets and achievements. The EMS you create should enable that reporting. How much documentation you choose to keep will no doubt reflect the type of reporting and communication that your organization uses to communicate with its stakeholders.

AREA	SHORT-TERM PROJECTS	LONG-TERM PROJECTS
Energy	Switch off lights	Fit motion sensors to light switches
Water	Fix Leaks	Convert to water saving fixtures
Waste	Stream your waste Reduce your waste paper by printing double-sided	Reduce your waste paper through changes in your work practices
Transport	Reduce the travel budget Training in teleconferencing facilities Set up a lift-share scheme	Provide increased teleconferencing facilities

FLEXIBILITY

This is a journey, a path, a striving to be green, and as you strive to be green you will encounter changes around you that will need to be absorbed into the EMS.

Your plans and your EMS should be flexible enough to accept these changes, and which will then help to anticipate where change will come from.

- The business structure will change. You will need to engage with different roles and people. This will affect how the champions operate and who they operate with.
- Legislation will be created. New legislation on where your organization is based and how it manufactures or delivers will all need to be responded to. Additional legislation is likely to require additional reporting, as with the CRC Energy Efficiency Scheme, or due diligence, as with the Lacey Act.
- Your understanding of best practice will change. Will well-managed forestry begin to assess carbon sequestration and require more set-aside or longer cutting cycles? Will certification of water use become the norm, and will your organizations and your suppliers need to demonstrate good water management through a widely accepted and structured approach? Reviewing targets and internal standards will help incorporate the new understanding of best practice.
- Your understanding of what constitutes "indirect impacts" that you can reasonably influence will evolve and expand. Reporting scopes as defined by the GHG

protocol require directly created impacts to be reported on, and ISO 14001 requires direct impacts and those within the organization's control to be reduced. There is a lot of work going into creating a reporting basis for indirect impacts and on what is in an organization's control. As these ideas develop, the requirements for reporting will expand, as will your ability to effectively manage your impacts.

- Supply chain management. The extent to which you can manage your supply chains will expand. How far down the supply chain should you take your responsibility, and what will constitute responsibility when you are working down a supply chain at several removes from your organization? As implied by the point above, how will this be accounted for? Will it be purely about management or will impacts also be accounted for through reporting? As an organization becomes more transparent, it will be better able to develop relationships of mutual responsibility with its suppliers, and they with their suppliers in turn.
- New products. As new products are created you will need to manage the impacts of potentially new manufacturing techniques and new materials. There will be challenges around accounting again. If producing information likely to be used as a set for comparison, for instance Lifecycle Analyses of a paper and a digital product, which elements of the lifecycle need to be included to produce results that are genuinely comparable? Content creation is comparable, but one product has no energy effects in

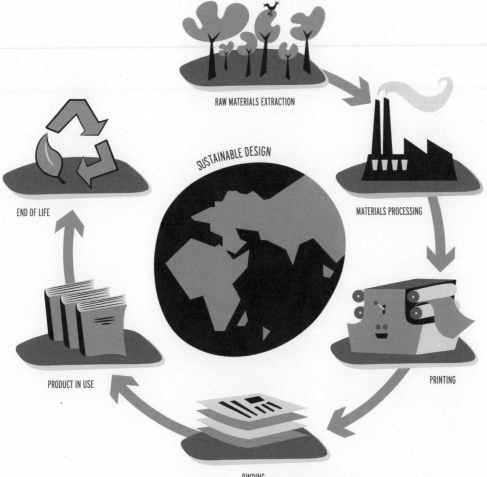

SUSTAINABLE DESIGN

RAW MATERIALS EXTRACTION

MATERIALS PROCESSING

PRINTING

BINDING

PRODUCT IN USE

END OF LIFE

use while the other does; for one the publisher is responsible for the manufacturing, for the other it is not.

• Manufacturing technologies will develop. Energy requirements will change. Will there be less or more waste? Again, this is complex for digital projects as you are not responsible for a key part of the supply chain that will get the product to the reader.

These last three points are where we see the inclusion of indirect impacts. Your manufacturing impacts can be managed by an EMS whether or not you choose to have it certified. The goal of the EMS and the structure will be the same for an organization's direct and indirect impacts, so they can be managed by the same process.

From this chapter we can see how the members of an organization can collectively begin to change that organization. We have looked at the areas where direct business impacts can be reduced and the many types of benefit this provides an organization. We have also seen how to create a productive and lasting EMS to manage the process of change, for direct and indirect impacts. In the next chapter we will start to look at how you find the environmental production workflow appropriate to your organization.

Above: Sustainable design and circular lifecycles.

Choosing & setting up an eco-friendly workflow: the green supply chain

So far we have been looking at the big picture and the wide range of macro-environmental issues which businesses, wanting to reduce the level of their impact on the environment, need to take into account as they develop their strategic plans and manage their day-to-day activities. These impacts are, broadly speaking, consumption of energy and raw materials, waste, and pollution. We have also been looking at how practice and attitudes can be, and are being, changed through global drivers, legislation and regulation, and frameworks for action. The aim of this chapter is to look at how businesses, and publishing in particular, are responding to the need for change, and at what they are doing to reduce the size of their environmental footprint.

The business environment

Business, whatever its aim, operates in an environment which is dynamic and complex. It is dynamic because it is in a state of constant change, and it is complex because it consists of a number of different but interrelated factors. These factors are political, economic, social, technological, legal, environmental/ecological, and ethical.

Until quite recently there were only four factors—political, economic, social, and technological—known by their acronyms: PEST or STEP. The addition of the legal, environmental, and ethical factors is an interesting sign of the times and is an indication that the business community is increasingly aware of the need to recognize the key part that environment and regulation play not only in their strategic thinking and how they operate, but in how they produce products and bring them to market. These factors are now known by the acronym STEEPLE.

These factors constantly change, and a company that wants to remain in business and grow by creating products that are relevant to its market needs to be aware of these changes, so that it can adapt its strategy and practices accordingly.

One way in which companies are doing this can best be seen in their move away from traditional supply chain management and their increasing adoption of green supply chain management, which has been prompted as much by enlightened self-interest as by a realization that environmental legislation and regulation, as well as consumer demand and expectation, are powerful incentives for doing things in a different, greener way.

THE TRADITIONAL SUPPLY CHAIN

If you look at a traditional supply chain model, you can see that it is made up of a chain of different organizations whose aim is to take various raw materials, bring them together, and transform them during manufacture into a product, which ends up in the hands of the consumer.

In the case of a book (or other printed products) the traditional supply chain looks like the diagram opposite.

Books are made from virtually 100% paper, and the basic raw material for this is wood pulp. Trees are extracted from the forest and sent to a pulp mill, where they are converted into pulp. (For a detailed description of this, and other paper-making processes, see Chapter 5.) The pulp is converted into paper at a paper mill, and from there it makes its way to a supplier, where it is printed, bound, and transformed into a book so that it can be delivered to a wholesaler/retailer for onward sale to the consumer. This supply chain also holds true for newspapers and magazines.

As can be seen in the diagram, each stage (or process) involves the transformation of an input into an output (or product), which then becomes the input for the next stage until manufacturing is completed and there is a final product. Once the product reaches the consumer the supply chain ends, having achieved its purpose.

In this supply chain model, once the product has been sold it becomes the responsibility of the consumer. What happens to it, and how it is disposed of once it has reached the end of its life, is of no concern to the producer.

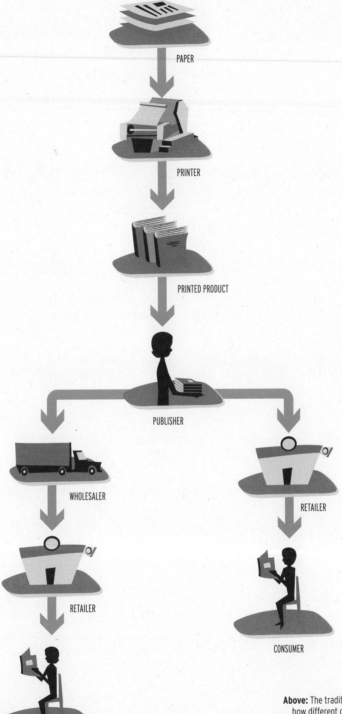

PAPER

PRINTER

PRINTED PRODUCT

PUBLISHER

WHOLESALER

RETAILER

RETAILER

CONSUMER

CONSUMER

Above: The traditional supply chain, showing how different organizations transform raw materials and outputs from different processes into the finished product, and get it into the hands of the consumer.

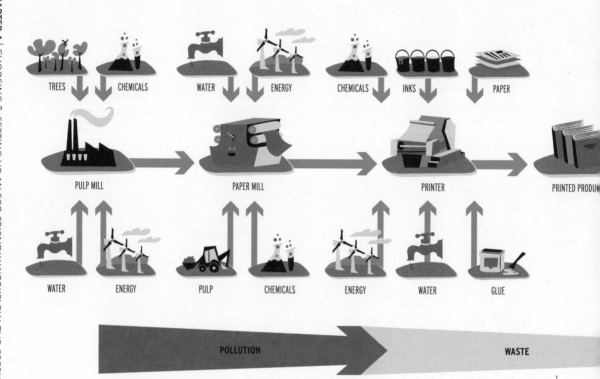

THE GREEN SUPPLY CHAIN

In the green supply chain model, things are different, and the producer's involvement with their product extends virtually from the cradle to the grave—from procurement to disposal—and involves producers in extended product responsibility (EPR).

In a green supply chain, the focus is extended to take into account a whole range of issues:

- depletion of natural resources
- climate change potential
- cumulative energy demand
- water consumption
- waste

In a green supply chain, the emphasis is on sustainability through working as closely as possible to the principles of the three Ds, and of the three Rs (or four Rs, if Recovery is included).

The principles of the three Ds are:

- **Dematerialization:** reducing the amount of materials used to make products (in the case of publishing this is predominantly paper)
- **Detoxification:** reducing, if not altogether eliminating, the use of substances harmful to health (for example, solvents in printing inks)
- **Decarbonization:** reducing the amount of energy used to make and deliver products (for publishing this could mean printing fewer copies, and printing closest to where the market is to reduce transportation and delivery costs)

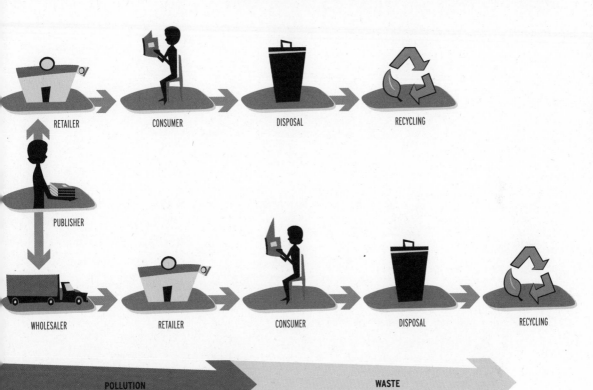

VALUE CONSUMPTION

VALUE RECOVERY

RETAILER

CONSUMER

DISPOSAL

RECYCLING

PUBLISHER

WHOLESALER

RETAILER

CONSUMER

DISPOSAL

RECYCLING

POLLUTION

WASTE

Above: The green supply chain showing how the producer's involvement with their product extends from the "cradle to the grave" as they transform raw materials into the end product. Energy, water use, pollution, and waste are all now factors which publishers need to take into account.

"In a green supply chain, the focus is extended to take into account a whole range of issues."

Right: The waste minimization hierarchy pyramid showing the options available for minimizing waste. At the top of the pyramid comes the most favored and effective option, **prevention**, followed by reduction, all the way down to the bottom to **disposal**, the least favored option. Prevention is better than cure.

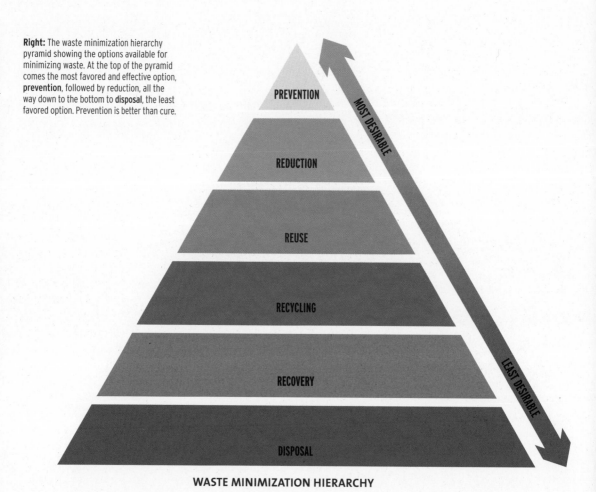

PREVENTION

REDUCTION

REUSE

RECYCLING

RECOVERY

DISPOSAL

MOST DESIRABLE

LEAST DESIRABLE

WASTE MINIMIZATION HIERARCHY

The principles of the three Rs are:
- **Reduction:** already covered under the three Ds
- **Reuse:** incorporating parts of the original product into a new product. This is something rarely done in publishing, mainly because the principal raw material is paper, which is easier to recycle than, say, lead from a car battery
- **Recycling:** processing used or waste material into a new product

The Rs form part of the waste minimization hierarchy.

At the top of the pyramid (in order of priority) comes prevention, followed immediately by reduction. It is easier to prevent waste than to treat it or clear it up afterwards. It is smarter to meet a need than exceed it.

A company that can apply these two principles to its products as well as to how it operates is likely to have a much smaller environmental impact than one that relies on the strategies of reuse, recycling, and recovery lower down the pyramid—with simple disposal being the least favored one at the bottom.

To start the move towards a green supply chain, a company needs first to become aware of the environmental aspects and impacts of how it operates, and how it makes and sells its products. Until it has done this, it is not really in the best position to know how to set about reducing its impacts.

The first step is for the company to carry out a Lifecycle Analysis (also known as a Lifecycle Assessment) of its activities and its products.

According to **ISO 14040(2006)** *Environmental management: Lifecycle assessment: Principles and framework*, and **ISO 14044(2006)** *Environmental management: Lifecycle assessment: Requirements and guidelines*, a Lifecycle Analysis should consist of four phases:
- goal definition and scoping
- inventory analysis
- impact assessment
- interpretation

By carrying out a Lifecycle Analysis a company can:
- identify where the raw materials it uses in its products come from, and how sustainable these natural resources are
- quantify how much energy is used in making its products
- quantify how much pollution and waste the manufacturing processes release into the environment
- identify how its products are packaged and distributed
- identify how its products are disposed of when they come to the end of their life
- assess the possible environmental effects of its activities and its products in terms of energy, water, and materials use, and pollution and waste releases

There is a wide range of tools which can be used to carry out a Lifecycle Analysis. Some of these are complex and suited to industries where the supply chain can be more complex and involves many suppliers working all over the world. But in publishing, a Lifecycle Analysis is fairly simple, since the products are relatively uncomplicated, unlike, for example, a motor car or a computer. Reflecting this, the supply chain is also relatively uncomplicated, and quite a large proportion of it is green already, making it possible to produce green products without too much trouble: paper can be procured from FSC or PEFC certificated sources, and there is an increasing number of suppliers (printers and binders) who have received FSC or PEFC chain of custody certification and are also able to print and bind in ways that are environmentally friendly. Publishers in the UK can now calculate their greenhouse gas (GHG) emissions by using the web-based carbon reporting tool, created by the British Booksellers' Association and Publishers' Association Environmental Action Group.

While pulp and paper makers, as well as other suppliers, make the move towards reducing their environmental impacts, there is still a significant number of publishers who have yet to engage in the process, both at organizational level and in the products they produce.

Becoming green as an organization is one part of an overall process, and although it is an important part, as has already been discussed in Chapter 2, it is not the whole picture: the products created by a green organization need to be green as well, and this can only happen if greenness is integrated into the products through green design at the start of the supply chain.

GREEN DESIGN

Producing products that are environmentally friendly, and producing them in an environmentally friendly way, doesn't happen by itself: it happens by design. Everything that is green about a product has to be thought about and decided on.

Books, magazines, and newspapers, unlike cars, are reasonably easy to produce and they have fewer variables. For these reasons it is easier to make them green. So how do we go about turning our products green? We need to start by changing the way we think about our products. We need to move away from seeing them just as printed objects, and see them as part of the supply chain instead, which starts with procurement and ends with disposal.

Taking a book as a case in point, if we want to design it so that it becomes green, we need to go beyond the graphic and typographic, or how content is going to appear on the page, and consider other factors, which may at first seem quite trivial: for example, the number of words the author has been asked to write has a direct effect on the number of pages in the book (extent), the amount of paper needed to print it, how heavy it will be (weight), and how thick it will be (bulk). These in turn have an effect on the supply chain that reaches upstream to the raw materials, and downstream to delivery, distribution, and disposal:

- How many trees were used to produce that paper?
- What quantity of chemicals, water, and energy were consumed in producing it?

- How much GHG, waste, and pollution was released into the environment when it was being made?
- How much packaging will be needed?
- How many truck loads will it take deliver the stock to the warehouse or distributor, and how much GHG and other pollution will this release?

Thus, other seemingly unimportant factors now become important:

Paper weight

(also known as the substance or grammage)
A 100gsm paper has more pulp in it than a paper that weighs 80gsm, so it has consumed more energy, chemicals, and water and caused more pollution than the 80gsm paper did. It also costs more, since paper is sold by the metric ton. A book printed on a 100gsm paper costs more to transport than its 80gsm equivalent, not only in terms of money but also in terms of its environmental price tag. Interestingly, paper does not necessarily become more opaque as it gets heavier: all that happens is that the book gets heavier, and the cost of paper goes up. A 70gsm paper can be just as opaque as a 100gsm one. Opacity is more a consequence of how a paper is made, and what goes into it and onto it, than weight alone.

Paper thickness

(also known as bulk, calliper, gauge, or volume)
Thick paper increases a book's volume, making it fatter or more bulky. Bulky books take up more space than thin

books. What might have been a single truck load becomes two, with all that brings in the form of emissions as books are transported from the supplier to the publisher's warehouse and from there to the consumer. A load that might have fit onto six pallets now needs nine, each one of which has to be moved into and out of the truck and into and out of the warehouse. The pallets are either metal or wood, and had to be made, using raw materials and energy, and creating pollution in the process.

Paper type

Paper comes in a wide range of types, from newsprint for newspapers to coated papers for high-quality color reproduction. As you will see in the next chapter on raw materials, newsprint is relatively economic in the demands it makes on the environment in terms of transport, trees, chemicals, water, energy, waste, and pollution. As paper quality increases, so do the demands: the higher the quality the greater the environmental price tag, with coated, acid-free (or permanent) paper at the top of the league. The most effective way to reduce these impacts is to use recycled paper wherever practicable.

So far we have been concentrating on paper, but there are other factors that need to be considered as being part of the green design treatment. For example, the use of color, and whether illustrations bleed. Color work usually requires a coated paper for the best result, and

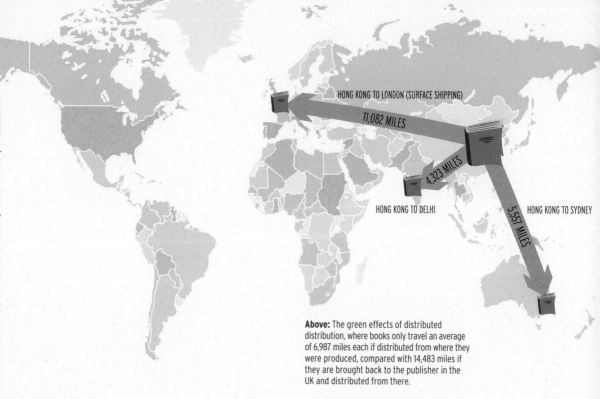

HONG KONG TO LONDON (SURFACE SHIPPING)

11,082 MILES

4,323 MILES

5,557 MILES

HONG KONG TO DELHI

HONG KONG TO SYDNEY

Above: The green effects of distributed distribution, where books only travel an average of 6,987 miles each if distributed from where they were produced, compared with 14,483 miles if they are brought back to the publisher in the UK and distributed from there.

as we know already, the impacts of coated are greater than those of uncoated paper.

Bleeding illustrations require a larger sheet of paper to allow the grippers on the printing press to pull the sheet through the press without coming into contact with the image area and to allow for trimming. This in turn uses more paper, which, in the end, is cut away from the book and disposed of as waste. If the binder has adopted green practices then that paper will go for recycling. But if the binder is not green, then that waste will more than likely end up in landfill, where it will steadily release its pollutants into the environment as it decomposes.

In Chapter 5 we will go into more detail about how to integrate green practices into design and prepress. But by now, it must be clear that to be able to design a green product implies an understanding and knowledge of the supply chain and raw materials; and to illustrate this it would be useful to use a case study of a recent publication to see how some of the principles of green design could be put into action, before we move into the downstream reaches of the green supply chain to look at suppliers and disposal. (See opposite.)

This case study illustrates the importance of thinking green during the design of a project. Yes, it does require greater organization at the outset to distribute stock from a printer than it does to bring stock back to base, repack it, and send it off on its travels again. It might also involve greater trust. But if, by doing this, both the financial and the environmental impacts of moving stock around the globe can be reduced, then it is certainly worth the effort.

We now move into the downstream reaches of the green supply chain and look at suppliers, manufacturing, and disposal. These topics are also dealt with in detail in Chapters 6, 7, and 8.

In a green supply chain all the links need to be as green as possible: everyone involved needs to think green, be green, and act green; this applies to printers, binders, warehousing, and transport.

When a job is placed with a green printer or binder, there should be an increased interaction, a synergy, a collaboration between the client and the supplier, who are both working to a common goal: creating a product that is green, having a reduced impact on the environment,

CASE STUDY

This case study has to do with a textbook of veterinary medicine for nurses on a three-year university training course, with sizeable markets in the UK, Australia, India, and Hong Kong.

Because the book covered all three years of the course it was quite long, with an extent of 1,088 pages. The format, A4 (8.27 x 11.69 inches), was also quite large. Because there were lots of color illustrations and diagrams appearing randomly throughout the book, one of the production manager's main concerns was to make sure that the paper was opaque enough to prevent the illustrations on one side of the sheet from being visible from the other (known as show-through). To achieve this, a 100gsm coated paper with a volume basis of 16 was chosen, which would have produced a book weighing 7.26 pounds, with a bulk of 3.43 inches. While the marketing department, when they learned of this, said that it would be hard to sell a book weighing this much and being this thick, there are also the environmental issues we have discussed which need to be considered:

• **raw materials**

• **energy**

• **water**

• **transport**

• **packaging**

• **waste**

• **pollution**

It was the overriding concern with opacity that had led to the initial decision to buy such a bulky and heavy paper, and something needed to be done to change this. In the end, a 70gsm coated paper with a volume basis of 16 was chosen, which reduced the weight to 5.28 pounds, and the bulk to 2.4 inches, and satisfied the criteria for opacity. A paper with a volume basis of 12 would have reduced the bulk to 1.8 inches.

The second part of the case study involves the decision to print in Hong Kong (for reasons of economy) and to ship stock back to the UK; from where it would be distributed to Hong Kong, India, and Australia.

In purely financial terms doing this is expensive: the publisher is paying to transport the books 11,082 miles from Hong Kong to London, where they are repacked, and then sent back out again to Hong Kong (11,082 miles away), Sydney (13,242 miles), and New Delhi (8,046 miles)–an overall total of 43,452 miles.

The average distance traveled by each book is 14,483 miles (11,082 miles from Hong Kong to London, plus the average of the distance between London and the three main markets). If this is expensive financially, it certainly is in terms of the environment. If the publisher had considered the environmental impact alone of all these book miles, they would probably have gone about this differently. They would have continued to print in Hong Kong, and would have then arranged with the printer to distribute the books from there to the main markets, in which case figures for the book miles would look like this: Hong Kong to Hong Kong 0 miles, Hong Kong to London 11,082 miles, Hong Kong to New Delhi 4,323 miles, and Hong Kong to Sydney 5,557 miles.

The average distance traveled by each book is now 6,987 miles miles, roughly six times less than the book miles clocked up by bringing them back to London. The financial savings of doing it this way are relatively high, but the environmental ones are even higher.

Right: In a green supply chain all the links need to be as green as possible: everyone in the chain—printers, binders, warehousing and transport companies—needs to think green, be green, and act green.

that meets green regulatory criteria and standards, and that satisfies consumer expectations.

So how does one know that one's suppliers are green and, if green, how green they are? The first thing to do is to ask the printer to provide you with information about their environmental credentials, their environmental policies, and their environmental performance; the second is to audit these against the green printer checklist that appears in Chapter 7. As you work your way through the list you will be able to judge for yourself how green your supplier is. As already mentioned, going green is a process and involves different shades ranging from dark to light, depending on where a person or organization is in that process. A dark green supplier may well get a check for every box on the list, whereas another may only manage half. Whether to use the dark green supplier in preference to the light green one depends on a lot of factors other than just enviromental. However, one important factor is that the supplier, whatever their shade of green, understands the client's green strategy and works with it as far as is possible. The fact that you are starting to think green, and are looking for green suppliers, is a significant first step in the right direction.

The same applies to the people you use to transport your books to the warehouse, or wholesale or retail outlets, and to the people who store the books and distribute and deliver them. Some are more green than others; some are not green at all. Clearly, changing from a non-green transport company to one that has a green policy and

can demonstrate this in its performance is a lot easier than changing warehousers and distributors. But if your company is to be part of a green supply chain and play its part in the reduction of its global impacts, then these changes need to be made.

The final stage in the supply chain is disposal. As we already know, when compared with a car or computer, both of which are made up out of hundreds and hundreds of different pieces, a book is really quite simple, being made virtually out of one raw material: paper. Once a book has been bought it might stay on the purchaser's shelf for years to come, even be passed down from generation to generation. Some books can gain value with age, and even those that don't have a tendency to linger on in secondhand bookstores and websites, and there is always a used book stall at a flea market. The truth is that people generally do not like to throw away or destroy books.

However, it is when books are not bought that disposal becomes an issue. All publishers hope that their publications will be successful and continue to sell for years to come, become classics, and run to several editions. Each publication is as carefully developed as possible to realize this potential, but it does not always happen. The market may be sluggish and no one seems to want the book—least of all the retailers, who after a few weeks of holding slow- or non-selling stock in their shops want to get it off their shelves as fast as they can and back to the publisher, who will (though this depends

"Going green is a process and involves different shades ranging from dark to light, depending on where a person or organization is in that process."

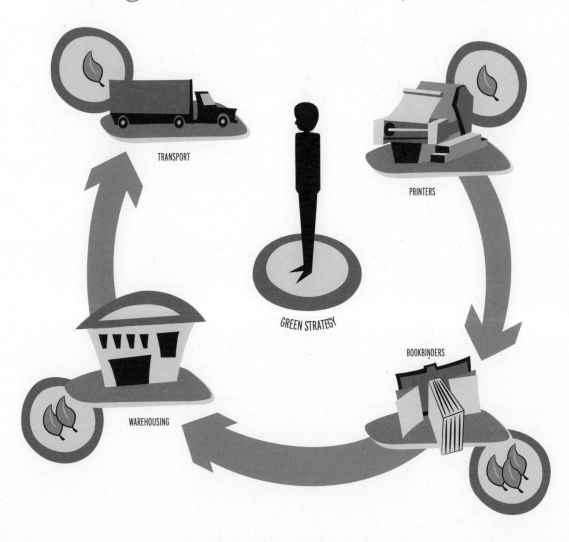

TRANSPORT

PRINTERS

GREEN STRATEGY

BOOKBINDERS

WAREHOUSING

on the terms of sale to the retailer) issue them with a credit note for the returned unsold stock. This system of returns appears to be unique to the publishing industry, and various estimates put the rate of returns at 30-40 percent of books sold. Some of this stock can be sold to remainder dealers, but a high percentage of returns (some estimates say between 65-95 percent) is pulped by publishers and sent to landfill, where it sits emitting GHGs into the atmosphere and ink and chemicals are leached into the ground.

Going back to the waste minimization pyramid, the best way to deal with waste is to prevent it from happening in the first place—disposal is a last resort, at the bottom of the pyramid.

One thing that publishers can do to reduce the waste from returns is to reduce their print runs and produce fewer copies. The technology is now there to make it possible to respond to how the market is behaving and to print copies in small numbers economically and quickly. No longer is it necessary to print 5,000 copies all at once in the hope that they will all sell. Now it is possible to print books little and often: for example, in ten batches of 500, or fifty batches of 100. Publishing in this way reduces a publisher's exposure to risk; it reduces inventory and money tied up in stock that won't sell. But most of all it reduces the impact on the environment of what publishers do and how they do it.

To end the chapter, here is a press release from the U.S.-based Book Industry Environmental Council, which appeared on April 16, 2009:

"Today, the Book Industry Environmental Council announced a goal of reducing the U.S. book industry's greenhouse gas emissions 20% by 2020 (from a 2006 baseline) with the intent of achieving an 80% reduction by 2050.

"This industry-wide commitment is a global first in publishing.

"In 2008, an industry-focused report concluded that the U.S. book industry has a climate impact equivalent to 12.4 million metric tons of carbon. Using the same methodology as this report, this 20% reduction will represent a savings of up to 2.5 million metric tons per year, the equivalent annual emissions of approximately 450,000 cars."

Three years later, it is probably a bit early to tell what effects this initiative has had, or will have, and whether that 2050 target will be achieved. But one thing you can be certain of: any contribution that you can make by thinking and acting green, and ensuring that your thinking and actions carry through to your supply chain, is going to make it a lot more likely that it will be achieved.

"The technology is now there to respond to how the market is behaving and to print in small numbers economically."

Left: The Book Industry Environmental Council announced a goal of reducing the industry's GHG emissions by 80% by 2050. By thinking green and acting green between now and then, we are much more likely to achieve this target than if we carry on as before.

How to ensure you buy green raw materials

In this chapter we are going to look at the raw materials you, or your suppliers, source to produce your products. We'll be covering paper and board, inks, glue (or adhesives), and cover finishes (lamination and varnishes). We shall start by looking at how these raw materials are produced and at how they are used in order to develop an understanding of the impact they have on the environment, before moving on to provide you with a range of strategies and options that will help you to reduce these impacts.

Paper & board

SOFTWOOD TREES

HARDWOOD TREES

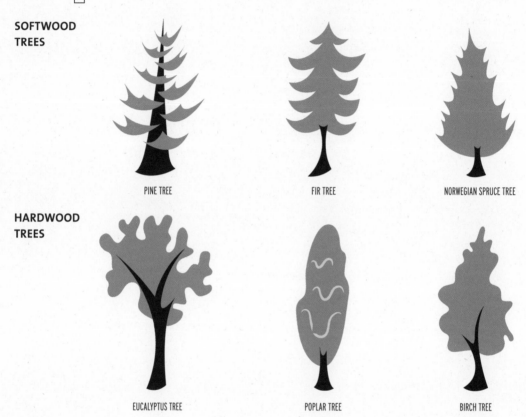

PINE TREE   FIR TREE   NORWEGIAN SPRUCE TREE

EUCALYPTUS TREE   POPLAR TREE   BIRCH TREE

The single largest component of any printed product is paper, and this is where we are going to begin. Since the mid-nineteenth century, paper has been made predominantly from trees—hardwood as well as softwood. Softwood trees (fir, pine, and spruce) produce long cellulose fibers that give paper its strength; hardwood trees (like poplar, birch, and eucalyptus) produce fibers that give paper its smoothness, bulk, and opacity.

Trees are grown commercially in plantations, and are harvested when they have reached their maximum timber potential, which in the case of Sitka spruce is forty to sixty years—relatively rapid when compared with oak, which can take up to 150 years or more. Once felled, much of the mature tree is sent off to be used in furniture-making or by the construction industry. Paper is made from what is left over: the off-cuts and trimmings, as well as from younger trees known as "thinnings," which are cleared from the plantation after twenty years or so.

Although plantations contain growing trees, which should be good, they are not like the ancient, natural

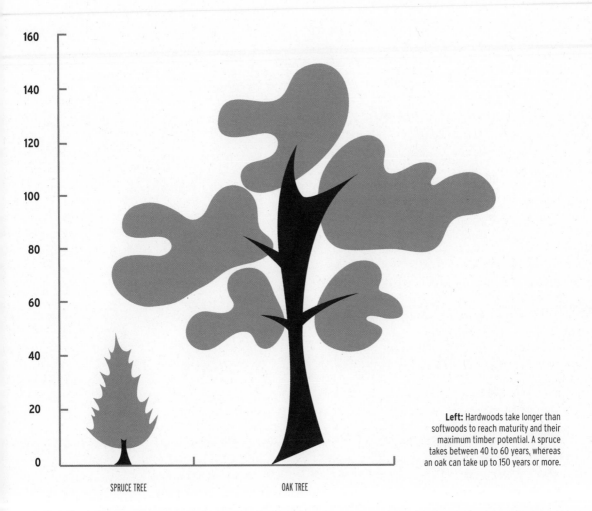

Left: Hardwoods take longer than softwoods to reach maturity and their maximum timber potential. A spruce takes between 40 to 60 years, whereas an oak can take up to 150 years or more.

SPRUCE TREE

OAK TREE

forests they increasingly replace. Where the ancient forest was marked by diversity and natural chaos, the plantation usually contains one kind of tree, uniformly spaced apart to maximize access to light and air. Natural nutrients have been replaced with pesticides and fertilizers. As the demand for paper pulp increases, so does the march of the plantations, taking with them all that lies in their path: ancient forests and the bio-diverse communities of people, animals, and plants that live in and depend on them.

Plantations and ancient forests need to coexist, and this relationship needs to be managed, which is where government and organizations like the Forest Stewardship Council (FSC) and the Program for the Endorsement of Forest Certification (PEFC) come in. Their role, which has already been discussed in detail in Chapter 1, is "to promote the responsible management of the world's forests" (FSC website), and "set standards and realistic criteria for sustainable forest management" (PEFC website). Despite efforts through legislation, and forest and chain of

PUBLISHERS' DATABASE FOR RESPONSIBLE ENVIRONMENTAL PAPER SOURCING
Inspired by the sharing of the Egmont Grading System ©

Egmont	Usborne	Cappelen Damm
Imago	Walker Books	Wissenmedia
Hachette UK	Cambridge University Press	McGraw Hill
HarperCollins	John Wiley & Sons	Scholastic
Pearson	Oxford University Press	Macmillan
Penguin Books	Simon & Schuster	Random House
Reed Elsevier	Parragon	Chronicle Books
Sage	Meld	

custody (CoC) certification, to reduce illegal logging, there is still a significant trade in illegally logged timber, which will only really stop when the market is able to identify wood and fiber from illegal sources and refuses to buy it. So, one of the first things you can do to make sure that the paper and board bought by you, or your supplier, is green is to check that it carries FSC or PEFC certification, and to refuse to buy it or use it if it doesn't. (To find out more about the rules and standards governing chain of custody certification, go to Chapter 7.)

To help develop an "understanding of responsible paper supply chains," a group of twenty-three leading British and international publishers have set up a database known as PREPS (Publishers' database for Responsible Environmental Paper Sourcing). According to the PREPS website the database provides publishers with "the technical specifications and details of the pulps and forest sources of the papers they use. It also holds data on CO_2 emissions and water use at the paper mill level.

"Based on the forest source information, each paper is awarded a grade of 1, 3, or 5 stars using the PREPS Grading System. This considers the country of origin of the wood fiber and how the forest sources have been managed.

"PREPS users are able to access this information and take it into account when making purchasing decisions."

PREPS runs two grading systems: one for grading forest sources, and grading is done to the criteria which appear on p. 94; the other is for grading paper, and is based on the grade awarded to the forest source. The paper grading criteria appear on p. 92, and the flow diagrams showing how PREPS grades both certified and uncertified papers appear on pp. 91 and 93.

You can see how useful they are in showing if the paper you are planning to buy is from certified sources, and how important they are in preventing you from buying paper that contains pulp from non-certified or illegal sources.

Above: PREPS, the Publishers' database for Responsible Environmental Paper Sourcing, is supported by 23 British and international publishers whose aim is to help develop an understanding of responsible paper supply chains.

Right: PREPS grading flow diagram for certified papers.

THE PREPS GRADING FLOW DIAGRAM

This flow diagram shows how PREPS grades **certified** papers.

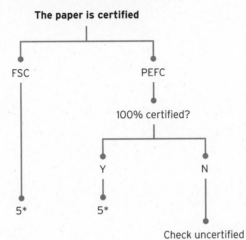

The grades in this diagram refer to the whole paper

The paper is certified

FSC

PEFC

100% certified?

Y

N

5*

5*

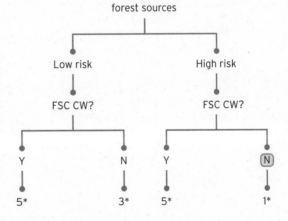

Check uncertified forest sources

Low risk

High risk

FSC CW?

FSC CW?

Y

N

Y

N

5*

3*

5*

1*

PREPS may request a paper sample to be fibre tested for papers that:

1: are manufactured in a country with a high trans-shipment or governance risk°

2: contain pulp(s) from a country with a high trans-shipment or governance risk°

3: contain forest sources highlighted in blue on the flow charts

° risk is determined using the Acona forest risk tool, which is available in the process document from the PREPS website **www.prepsgroup.com**

CW = controlled wood

Illustrations of the grading systems reproduced by kind permission of PREPS.

THE PREPS PAPER GRADING CRITERIA

GRADE	CRITERIA
***** Recycled, FSC or 100% PEFC certified	Awarded if: • the paper is certified and labelled FSC 100%, FSC Mix or FSC Recycled; OR • the paper content is 100% recycled; OR • the paper is entirely made of a combination of recycled, FSC and PEFC certified material; OR • for PEFC certified papers all the forest sources are from a certified source i.e. the non certified portion must be made up of FSC and/or FSC Controlled Wood forest sources.
*** Known and Responsible	Awarded if: • all material comes from a low risk source, as defined by the Country Forest Risk Tool, and is not originating from within a WWF-defined Ecoregion, OR • all high risk material is verified by either FSC or PEFC but the paper is not certified, OR • there is a PEFC certified paper with material originating from low risk to uncertified sources. If a forest source is high risk, or is from within a WWF Ecoregion, the source must be *verified*. *Verified* means that a source is either: • certified as meeting the FSC Controlled Wood Standard (FSC-STD-30-010); AND/OR • certified by a recognised forest certification scheme.
* Unknown or unwanted material	• Awarded if: • any portion of the paper is known to, or suspected to, come from a high risk source; OR • any of the material is from an unknown source. Material from a controversial source, which is not certified under an acceptable forest certification scheme, would also be considered high risk.

GRADING OF PAPERS

Papers are awarded a grading of 1, 3 or 5 stars based on a system knows as the PREPS Grading System. Each forest source is graded according to the System. These are then used to award the whole paper a grade (see table above: the PREPS paper grading criteria). These are regularly reviewed and amended to keep pace with developments in forest sourcing.

THE PREPS GRADING FLOW DIAGRAM

This flow diagram shows how PREPS grades **uncertified** papers.

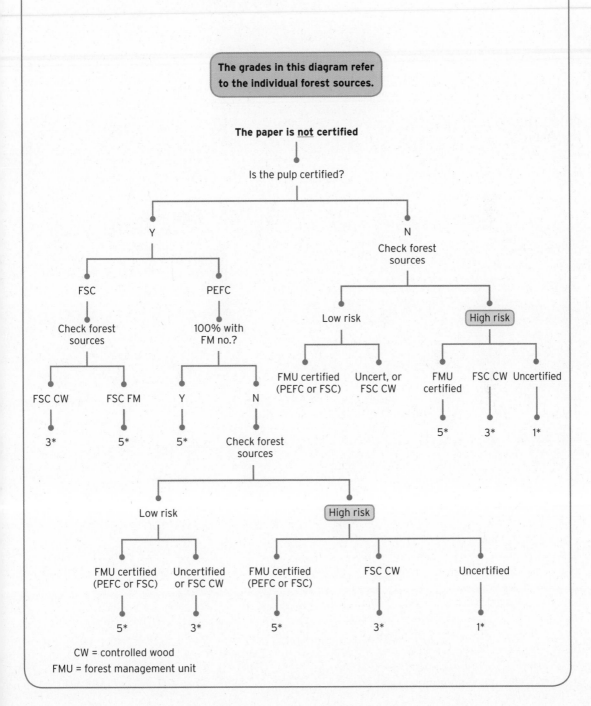

The grades in this diagram refer to the individual forest sources.

The paper is <u>not</u> certified

Is the pulp certified?

Y

FSC

Check forest sources

FSC CW

3*

FSC FM

5*

PEFC

100% with FM no.?

Y

5*

N

Check forest sources

Low risk

FMU certified (PEFC or FSC)

5*

Uncertified or FSC CW

3*

High risk

FMU certified (PEFC or FSC)

5*

FSC CW

3*

Uncertified

1*

N

Check forest sources

Low risk

FMU certified (PEFC or FSC)

Uncert, or FSC CW

High risk

FMU certified

5*

FSC CW

3*

Uncertified

1*

CW = controlled wood
FMU = forest management unit

FOREST SOURCE GRADING CRITERIA

GRADE	CRITERIA
***** **Recycled, FSC or PEFC certified**	**Awarded if:** • the forest source is certified by an FSC or PEFC Forest Management licence **OR** • the forest source is made up of 100% recycled material
*** **Known and responsible**	**Awarded if:** • the forest source material comes from a low risk country, as defined by the Country Forest Risk Tool, and does not originate from within a WWF-defined Ecoregion
* **Unknown or unwanted material**	**Awarded if:** • any portion of the paper is known to, or suspected to, come from a high risk source **OR** • any of the material is from an unknown source Unwanted material is material from a controversial source which is not certified under an acceptable forest certification schem

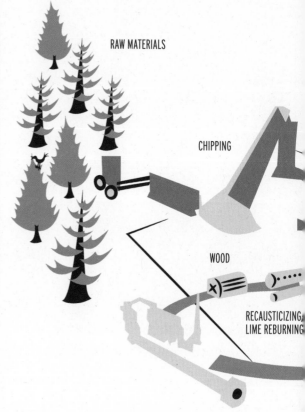

RAW MATERIALS

CHIPPING

WOOD

RECAUSTICIZING
LIME REBURNING

To gain full access to the database and take advantage of all that it offers you, you need to be a member of PREPS. But just being aware that such a system exists should alert you to the fact that some papers are more environmentally friendly than others, that you need to know which these papers are, and that by using only those papers that meet the PREPS 5- or 3-star paper and forest sourcing criteria, you are reducing the size of your environmental footprint and the size of the market for papers coming from high risk or unknown sources.

Once the trees in a plantation have reached their maximum size, they are cut down (or logged). Logging usually involves clear-cutting (or clear-felling), which involves removing all the trees in the plantation and is effectively a form of deforestation. The effects of clear-cutting are loss of habitat, increased water run-off and risk of flooding, and landslides and soil erosion.

In addition, roads need to be driven through the plantation so that trucks can remove the felled trees and take them to their final destination: a saw mill, where they are converted to timber, or a pulp mill if they are to become paper. These roads add to the problems of soil erosion and run-off, and the trucks that move along them release exhaust into the atmosphere containing fine particles,

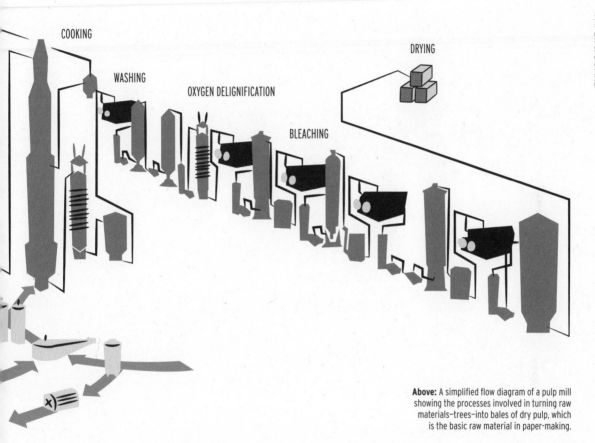

COOKING

DRYING

WASHING

OXYGEN DELIGNIFICATION

BLEACHING

Above: A simplified flow diagram of a pulp mill showing the processes involved in turning raw materials—trees—into bales of dry pulp, which is the basic raw material in paper-making.

soot, and noxious gases, which can include carbon monoxide when the engine is working under a load.

Pulp mills can exist on their own, producing pulp which is dried, baled, and sold on to paper mills. A paper mill that also produces its own pulp is known as an integrated paper mill.

The first stage in producing paper is to convert the wood into pulp, which is the basic ingredient used in paper-making. This process starts with debarking the tree and reducing it to small chips about 1-inch square. Turning these into pulp can be done using any one of the following processes:

- grinding the chips down between giant grindstones to produce ground wood pulp (SGW)
- crushing the chips between refiner plates using high pressure steam and heat to produce thermomechanical pulp (TMP)
- adding chemicals to the process, which separate the cellulose fibers (which are wanted) from the rest of the wood (lignin and hemicelluloses, which are not wanted). This kind of pulp is known as chemithermomechanical (CTMP) pulp

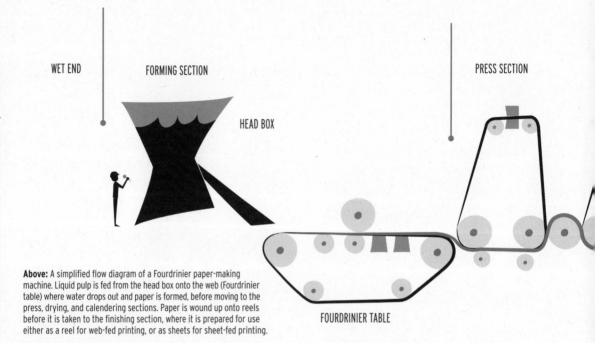

WET END   FORMING SECTION   HEAD BOX   PRESS SECTION   FOURDRINIER TABLE

Above: A simplified flow diagram of a Fourdrinier paper-making machine. Liquid pulp is fed from the head box onto the web (Fourdrinier table) where water drops out and paper is formed, before moving to the press, drying, and calendering sections. Paper is wound up onto reels before it is taken to the finishing section, where it is prepared for use either as a reel for web-fed printing, or as sheets for sheet-fed printing.

These three processes produce a pulp known generically as mechanical pulp, which is used mainly for newsprint or as a basic ingredient for lower quality papers. However, because it still contains much of what makes a tree a tree—lignin and hemicelluloses—papers made from mechanical pulp tend to discolor quickly, and they are not very strong. Nevertheless, on the plus side they are bulky and have good opacity, and the ratio of yield to trees consumed is very high, which means that it takes fewer trees to produce a ton of mechanical pulp than it does to produce a ton of chemical pulp, as can be seen in the table below:

PULP TYPE	YIELD
Ground wood and pressurized ground wood	95%+
Refiner and thermomechanical pulp	85-95%
Chemithermomechanical pulp	65-85%
Chemical pulp	45-65%

As the figures show, nearly three times as many trees are needed to produce a chemical pulp than a mechanical pulp.

At the opposite end of the scale lies chemical or kraft pulp. To produce chemical pulp, the wood chips are put into a digester where they are cooked. The combination of heat, pressure, and chemicals (sodium hydroxide and sodium sulphide in the kraft process) acts to remove the lignin and hemicelluloses. At the end of the process, the pressure in the digester is released and the wood chips explode into fluffy fibers resembling cotton wool.

Chemical pulps have longer fibers than mechanical pulps, because they are not smashed up or ground; they tend to be brighter, and they are also stronger. However, they are more expensive in terms of the chemicals used to make them, and they consume more trees. So, they have a higher impact on the environment than mechanical pulps.

The final stage of the pulp-making process is bleaching and washing. Bleaching increases the pulp's whiteness and decreases its natural woody color. Washing removes the chemicals.

Mechanical pulp is bleached using alkaline hydrogen peroxide (for whiteness) and sodium dithionite (for brightness), both of which have a relatively low impact.

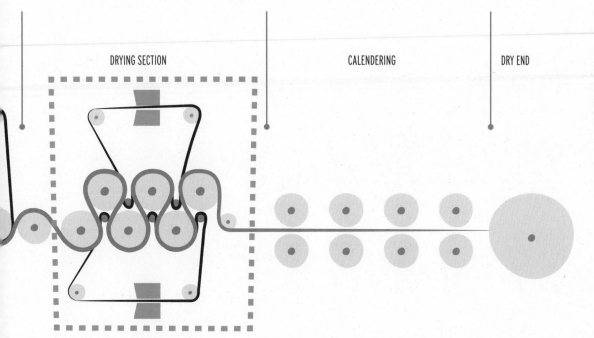

DRYING SECTION   CALENDERING   DRY END

Bleaching chemical pulp used to be done by using elemental chlorine gas. However, this method released so many toxic compounds into the atmosphere and waterways that it has been replaced by bleaching methods that do not use elemental chlorine. The first method—ECF (elemental chlorine free)—uses chlorine dioxide, and the second—TCF (totally chlorine free)—avoids the use of chlorine altogether, preferring to treat the pulp in a series of processes using oxygen, ozone, sodium hydroxide, alkaline peroxide, and sodium dithionite.

Once the pulp has been washed, it is dried and baled and is ready to be made into paper. This is done at a paper mill, which can either be fully integrated, in which case it produces its own pulp and makes paper from it, or non-integrated, which means that it buys pulp from a pulp mill and turns that into paper.

Although pulp is the basic ingredient of paper, additives, known as fillers, are added to it as it is being mixed into a slurry with water prior to being formed into paper on the Fourdrinier paper-making machine.

Fillers serve two purposes: the first is to reduce the amount of wood needed to make the paper. The second is to improve the properties of the paper and its performance on press. Fillers can be calcium carbonate (chalk) for increased whiteness and opacity, kaolin (china clay) for greater smoothness and gloss (especially when polished or calendered), or titanium dioxide for extra brightness and opacity.

Size (or sizing), either as glue or starch, is also added to the pulp at this stage to reduce its water absorption properties. Any pigments needed for coloring will be added as well.

The pulp is now ready to start its journey to becoming paper, which takes about three to four seconds from the moment that it leaves the head box as slurry at the wet end of the paper-making machine to its arrival, as a wet strip of what is now identifiably paper, at the press section, where any excess water is removed. From the press section the paper moves to the dryer section, where the moisture content is reduced further, before it is reeled up and removed for conversion into smaller reels or sheets.

An average paper mill produces about thirty tons of paper an hour, or about 250,000 tons of paper a year.

1 TON OF UNCOATED PAPER FROM CHEMICAL PULP TAKES 23 TREES

1 TON OF UNCOATED PAPER FROM MECHANICAL PULP TAKES 12 TREES

1 TON OF COATED MAGAZINE PAPER MADE FROM SEMI-MECHANICAL PULP (A BLEND OF MECHANICAL AND CHEMICAL PULPS) TAKES 15 TREES

1 TON OF COATED MAGAZINE PAPER MADE FROM MECHANICAL PULP TAKES 8 TREES

Quite how many trees it takes to make a ton of paper is difficult to work out since there are a lot of variables that have to be taken into account, such as the size of the tree in terms of its height and diameter, the kind of pulp it is made from (mechanical or chemical), and whether it is coated or uncoated.

Despite this, there are some helpful indicators that can be used to give a rough idea of the number of trees needed to produce 250,000 tons of paper, where the tree in each case is forty feet high and has a diameter of six to eight inches, which is the optimum average for trees that are going to be turned into paper. (See the chart above for examples.)

On top of the trees and chemicals that go into paper making, we need to think of other inputs, such as energy and water, and we also need to think of the outputs apart from paper (such as waste).

To cut down on energy use, mills use the bark and trimmings from trees, as well as the black liquor which is a by-product of producing chemical pulp. As might be

"An average paper mill produces about thirty tons of paper an hour, or about 250,000 tons of paper a year."

expected, chemical pulp used to produce fine and coated papers takes more energy than mechanical pulp used to produce newsprint.

Water is an important ingredient in paper making, as it is used to wash the pulp after bleaching and to convert the dry pulp into slurry. At one time water consumption was prodigious, running to some 132,000 gallons per ton, with its often-untreated corresponding waste being released into the waterways.

Over the past forty years paper mills, in the USA and Europe, have started to bring in closed-water systems, which make it possible to clean and re-use the water used in paper making. The effect of the closed-water systems has been to reduce the amount of fresh water needed in paper making from 132,000 gallons per ton to below 2,640 gallons per ton, with a corresponding reduction in the amount of waste water effluent to virtually zero.

For paper mills that do not have closed water systems, the problem remains of how to deal effectively with waste. Most mills are able to do this satisfactorily by treating it themselves and meet government regulation on what is an acceptable level. However, there are some mills that are more successful at this than others. Again, the fine and coated papers, because of their low yields, produce more waste (7,930–13,200 gallons per ton) than newsprint (2,640–6,600 gallons per ton).

By now it must be clear that producing paper using virgin fibers creates a quite significant impact on the environment, and that this will continue for as long as we continue to use this kind of paper. But, short of not using any paper at all and moving entirely into e-publishing, there is an option that makes it possible to produce print on paper products with a relatively low environmental impact: use recycled paper.

Recycled paper & board

The UK's National Association of Paper Merchants (NAPM) website defines recycled paper as "paper that contains fiber from waste paper." This fiber comes from two categories:

- **pre-consumer waste:** paper that has left the mill but has not been used (for example, bindery trimmings)
- **post-consumer waste:** paper that has been used (for example, old newspapers (ONP), office waste, and residential mixed paper (RMP)).

Paper made with recycled fibers should not be confused with that made from recovered fibers, which come from paper that never left the mill (known as mill broke or internal mill waste), and therefore does not strictly qualify as recycled.

Recycled papers, like virgin papers, come with a wide range of claims about how green they are, and while it may be argued that using any recycled paper is better than using none, the gold standard for recycled papers is the one that carries the FSC Recycled label. This guarantees that the paper is made from 100% recycled fiber, at least 85% of which is from post-consumer waste, with the remaining 15% coming from verifiable pre-consumer waste. The only factor that FSC cannot guarantee is that the fibers originally came from an FSC-certified forest, which is a small price to pay given the environmental gains that come from using recycled paper: no new trees had to be felled to produce this paper, and the product has not ended its life as waste in a landfill site, leaching chemicals and ink into the soil, and producing green house gases (GHG), in particular methane, which according to the Green Press Initiative website is a greenhouse gas with "21 times the heat trapping power of carbon dioxide."

However, demand for recycled paper often outstrips supply. When this happens it may be difficult to source paper that meets FSC criteria. But there are a number of recycled certification schemes besides FSC, run by PEFC and NAPM. The PEFC scheme certifies recycled paper with a recycled fiber content of at least 70% that comes from PEFC-certified forests and wood from controlled sources.

The NAPM scheme concentrates on the immediate source of the recycled fibers whatever their origins. The only criterion is that NAPM-certified recycled paper must be "manufactured from a minimum of 50%, 75%, or 100%

PEFC/16-44-1485
PROMOTING SUSTAINABLE FOREST MANAGEMENT

"This guarantees that the paper is made from 100% recycled fiber, at least 85% of which is from post-consumer waste, with the remaining 15% coming from verifiable pre-consumer waste."

genuine paper and board waste fiber, no part of which should contain mill-produced waste."

Printers, generally speaking, are willing to work with, and supply, recycled papers, particularly since paper quality and performance have improved since the early days of recycling. Recycled papers today can be every bit as bright and as white as paper produced from virgin fibers; their opacity is high, and their smoothness means that they run well on press.

If you, and your suppliers, start the move towards producing your products on recycled paper and boards as described in Chapter 7, you will start the important process of greening your production practices and reducing the size of your environmental footprint. However, paper and board are only the beginning of this process. There is a lot more that needs to be done with raw materials such as ink, adhesives, and laminates to further reduce your footprint.

Inks

Until quite recently the only printing inks available were those based on mineral oils, and these have been in use for centuries. Their high degree of tack, their high concentration of pigments, and their ability to dry quickly have made them particularly suitable for high-speed offset (sheet-fed as well as web) as well as flexographic and gravure printing. Mineral oil- (or petroleum-) based inks have been—and will continue to be—used to print your products.

However, these inks have a significant impact on the environment, because they emit volatile organic compounds (VOCs) that cause air pollution and reduce the quality of the air we breathe (which is not much fun if you happen to have respiratory problems such as asthma or bronchitis), and they contain non-renewable ingredients such as petroleum-based resins, barium, copper, titanium, manganese, cobalt, and zinc, which have to be extracted from the earth. As we shall see later in this section, there are viable alternatives to these inks.

Whatever their source, inks need to have a set of basic characteristics that make them suitable for use on high-speed printing presses. These characteristics can be summarized as:

- a high degree of tack (or stickiness) so that they adhere to the paper during printing
- a high concentration of pigment to produce a strong color
- an ability to dry quickly so they do not smudge when handled, or set off onto sheets of paper stacked above and below

These characteristics come from three basic components: pigment, vehicle, and additives. The pigment gives ink its defining color and opacity. The vehicle, besides giving ink its viscosity and flow, holds the pigment in suspension and binds it to the paper. The ingredients in the vehicle vary according to the printing method and the type of paper. The additives, which are mainly chemical, act as:

- driers, for accelerated drying (for example, salts or soaps of cobalt, manganese, zirconium, vanadium)
- extenders, for increased coverage (for example, barium sulphate)
- anti-oxidants, to prevent the ink from drying out on the press and forming a skin (for example, hydroquinone)

The amount of each these ingredients varies to produce an ink suited to the printing method and paper it is being printed on.

"Whatever their source, inks need to have a set of basic characteristics that make them suitable for use on high-speed printing presses."

INK TYPE	DRYING PROCESS	PERCENTAGE OF VOCS IN INK
Paste		
Sheet-fed offset	Oxidation	0-20
Web offset (heat set)	Evaporation	34-45
Web offset (cold set)	Substrate absorption	2-20
Liquid		
Flexo-gravure solvent	Evaporation	40-70
Flexo-gravure water	Evaporation	0-2
Publication gravure	Evaporation	40-70

Left: Inks come in paste or liquid form, depending on the use to which they will be put. The table shows the different percentages of volatile organic compounds (VOCs) associated with each type of ink. VOCs are released into the atmosphere during printing, and affect the quality of the air we breathe, which is particularly problematic if you have respiratory problems like asthma, for example.

INK DRYING

Once the ink has been transferred to the paper it has to cure (harden) and then dry. How quickly this happens depends on the formulation of the ink, drying process, and nature of the paper. Ink can dry in a number of ways:

- **Absorption (or penetration)**
  Ink penetrates the paper fibers and is absorbed by them. The depth of penetration determines the speed with which the ink dries. This is a relatively slow process and is mainly used in cold-set offset printing.
- **Oxidation**
  Oxygen in the air combines with the vehicle to convert it from its liquid into a solid state. The process is accelerated by a catalyst, so that the ink dries quickly. This method is mainly used in sheet-fed offset printing.
- **Evaporation**
  Solvents in the vehicle are evaporated by heat to bind the pigment to the substrate. This method is used in gravure printing, but can also be used in heat-set web offset printing.
- **Polymerization, or radiation**
  Used with ultraviolet (UV) inks, which harden when exposed to radiation from UV light, or electron beam (EB) and heat from infra-red (IR) heaters.

Inks come in paste form for offset printing and look rather like decorator's paint. Inks for flexographic, gravure, and ink-jet printing are liquid. The VOC levels present in each ink depend on its type, although, as can be seen in the table above, liquid inks require a higher percentage of VOCs in order to function and dry properly.

To reduce the harmful impact of printing inks, ink manufacturers and printers have invested significantly to:

- minimize or reduce the levels of VOCs in ink
- minimize, reduce, or eliminate emissions of VOCs and other hazardous air and water pollutants
- develop and use bio-derived renewable raw materials from sources like soy, safflower, cotton seed, or canola to produce vegetable-oil based, or bio-renewable, inks
- develop inks for waterless litho printing

Many printing presses are now fitted with capture and control systems designed to reduce and recycle VOC emissions, and vegetable inks are used extensively in the industry from cold-set newspaper and magazine printing to heat-set web offset and quick setting sheet-fed offset.

Although their VOC emission rate is negligible, vegetable inks, like their petroleum-based equivalents, still have to meet the demands of high-speed printing.

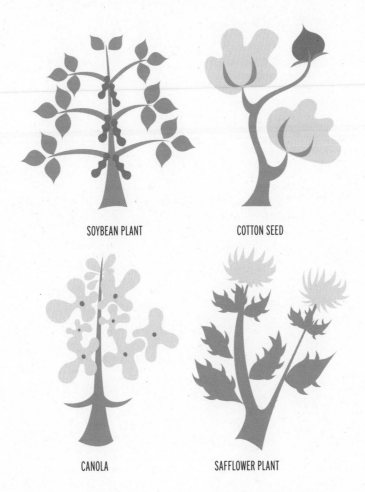

SOYBEAN PLANT

COTTON SEED

CANOLA

SAFFLOWER PLANT

For this, they need to have high tack, dry quickly, and deliver strong colors; this means that they contain virtually the same pigments and additives as petroleum-based inks, and although they are becoming more widely accepted, vegetable inks do take longer to dry than oil-based ones, because they contain little (if any) VOC emitting solvents, and this makes them difficult to use on coated stock.

But printing inks, whatever their source, like any other manufactured product, have an impact on the planet before, during, and after manufacture. As mentioned, oil-based inks contain non-renewable ingredients—minerals and pigments—that have to be dug out of the earth, or they contain vegetable oils from sources that have to be planted and harvested, sometimes at the expense of the surrounding natural habitat, as has been the case with soy production in the Amazon rainforest.

It is certainly worth talking to your suppliers about the possibility of using vegetable inks wherever possible, especially if you, or your company, are trying to reduce the size of your carbon footprint. Waterless litho is also a possibility, but unless your supplier is already set up as a waterless litho printer, it will require quite a lot of adaptation on their part not only to their presses, but to the environment in which they print, which needs to be air-conditioned for an acceptable standard of printing.

As we shall see in Chapter 7, buying green raw materials is one way of reducing your planetary foot print. Another way might be to look at your products and how you produce them in the first place.

Glues (adhesives)

Unless you are planning to saddle-stitch your books, they are going to contain glue, whether they are paperback or hardback (cased). In a paperback book, the cover needs to be attached to the spine, which is done with glue, and in a cased book, the case is attached to the book block by gluing it to the endpapers.

Adhesives can be classed as three types, each one based on the chemistry of different kinds of polymers. The first is made up of water-based emulsions (also known as cold-melts) and is based on polyvinyl acetate (PVAc) homopolymers, or vinyl acetate ethylene (VAE) copolymers. The second, known as hot melts, are based on ethylene vinyl acetate (EVA) copolymers, and the third, reactive hot-melts (also known as warm melts), are based on solid pre-polymer polyurethanes or PURs.

Modern adhesives, used in high-speed binding lines, need to dry quickly and to bond strongly. In addition, they need to allow the book to open easily, lie flat when open, maintain its shape, be flexible, be durable, and withstand stress and long periods in storage.

Although the water-based (PVAc and VAE) adhesives have their uses, they are relatively slow to dry, they have a relatively low page-pull strength, and they do not work successfully with coated stock.

Hot melt (EVA) adhesives dry and bond quickly, and can be used with any kind of paper. However, the high temperatures at which they are applied can cause problems with spine warping, especially if you are using cross-grained paper.

PUR adhesives, which are a relatively recent development, are increasingly used in paperback binding. They are more expensive than EVA hot melts, but need half the amount of glue to produce the same result.

All these adhesives are synthetic and contain chemicals that have an impact on the environment in one way or another, even though, by comparison with other raw materials, this may be fairly minor. For example, although PUR glues are solvent-free, like EVA glues they emit monomers, which in large quantities over a long period of time may cause health problems.

If books, magazines, and newspapers at the end of their lifecycle are recycled, glues, like inks, can be removed and safely disposed of—which is a strong argument for recycling. Where they are not recycled, and end up in a landfill site, then glues, like inks, will eventually detach themselves from the paper, leach into the soil and, inevitably, over time, find their way into the waterways.

From the supplier's point of view, there are no real alternatives: the only way they can be sure their books will hold together once bound is to use one of these glues.

For the supplier, and the publisher, the real issue is to work towards minimizing their impact on the environment even if they cannot eliminate it altogether. The skill is to achieve a balance between the demands of the customer and the needs of the environment: using a chemically based glue that emits a few monomers is preferable on all counts—economic as well as environmental—to having to reprint a run because the books have fallen to pieces.

HARDBACK

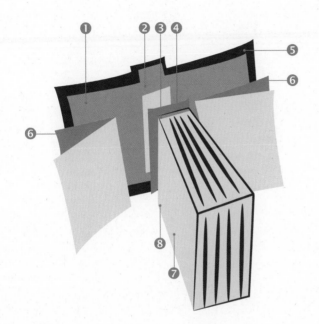

1. Board
2. Spine
3. Headband
4. Flexilining
5. Cover material
6. Endpaper
7. Section (book block)
8. Tailband (hidden)

PAPERBACK

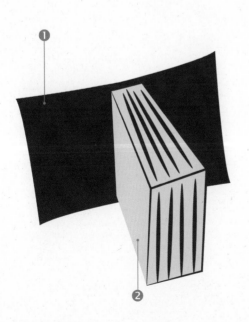

1. Cover
2. Section (book block)

Above: In hardback binding the case and the book block are held together by the flexilining, and by gluing the front and back endpapers to the boards of the case. In paperback binding the book block is attached to the spine by glue alone. Modern glues ensure that paperback books are durable and stay in one piece.

Lamination & cover/jacket finishing

Almost all covers and jackets are finished either with a film lamination or a varnish to protect them from scuffing and wear and tear, and to make them look attractive while on display in the bookshop. The most common form of lamination is either matte or gloss oriented polypropylene (OPP), usually 12-microns thick. Varnishes, which provide less strength than lamination, are either ultraviolet (UV) or the cheaper nitrocellulose (NC).

Like glues, laminates and varnishes are chemical-based, and like glues there aren't really any environmentally friendly alternatives, unless you count corn-based starch lamination as one. However, binders report that there are problems with it lifting from the cover or jacket after a while, which, in the long run, could be more environmentally damaging—if it resulted in a reprint—than if you had used a chemical-based product in the first place.

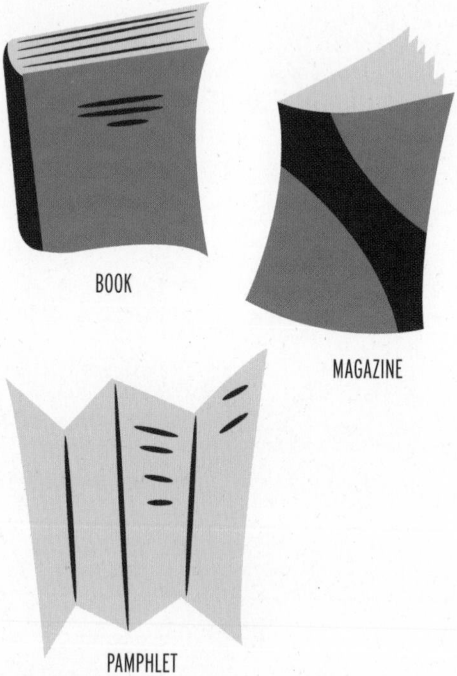

BOOK

MAGAZINE

PAMPHLET

We have now reached the end of the chapter on raw materials, and how to tell what is environmentally friendly and what is not. How this all adds up to a coherent picture and can lead to a production policy, if not an actual strategy, is the subject of the next chapter.

Above: Printed products of all kinds look better if they are finished with a sealant—either a varnish or film lamination—which enhances visual quality. At the same time it stops the ink from being scuffed, from showing signs of usage, and transferring from the printed surface to the end-user's fingers.

"Almost all covers and jackets are finished either with a film lamination or a varnish to protect them from scuffing and wear and tear, and to make them look attractive while on display in the bookshop."

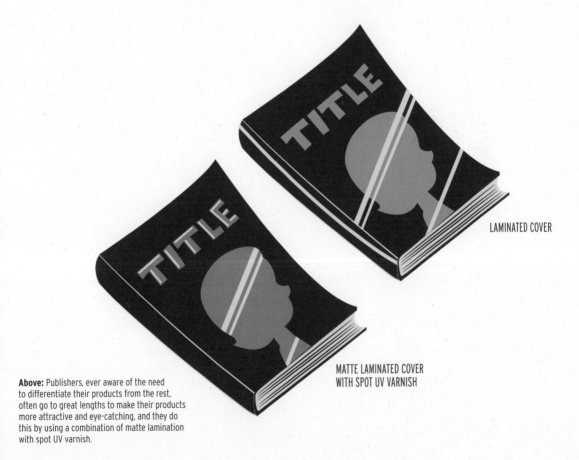

LAMINATED COVER

MATTE LAMINATED COVER
WITH SPOT UV VARNISH

Above: Publishers, ever aware of the need to differentiate their products from the rest, often go to great lengths to make their products more attractive and eye-catching, and they do this by using a combination of matte lamination with spot UV varnish.

Integrating green practices into prepress: decisions & design

The main focus of this chapter, and of the two that follow, is on how to create products that have a significantly reduced, if any, impact on the environment. To achieve this involves integrating the principles of the three Ds— dematerialization, detoxification, and decarbonization (or de-energization)—and the three Rs—reduction, re-use, and recycling—into how your products are designed, created, developed, and manufactured.

Design

Below: Different types of publication present different green design challenges.

NOVEL

INTEGRATED GUIDE BOOK

MAGAZINE

CHILDREN'S PICTURE BOOK

Design does not happen by accident: it is made up of a series of deliberate decisions about raw materials, manufacturing processes, and the look, feel, and function of the product.

We shall start by looking at the decisions that need to be made about the treatment of text and illustrations and their impact on the visual and physical appearance of the book, as well as on the environment, before moving on to deal with the decisions about format, paper, proofing, and printing technology.

TYPOGRAPHIC DESIGN: TEXT AND ILLUSTRATIONS

Typographic design is concerned with the presentation of text and (where they occur) illustrations on a page, as well as how these two elements relate to each other.

Opposite: The same text set in two different typefaces (Times New Roman and ITC New Baskerville) and point sizes (10 pt on 12 and 12 pt on 13), showing the difference in appearance between the two faces (fonts) and how extra space between the lines (known as leading) changes the appearance of the text, and its readability.

Vellacc usciusto temodi volupti sunt, utemolutempe ipsus, torit autem venda quiatem sam aceatem et explias aspiscil esequi ius arum comnist emquam, amusda coria comnit et alis ea quae. Nam volupta ssincium quatemporum hilis plaboreium voluptatia ne pos dolenectur, commole stibus a quiam venis qui natur, que quis sunt.

10/12 ITC New Baskerville Roman Std

Vellacc usciusto temodi volupti sunt, utemolutempe ipsus, torit autem venda quiatem sam aceatem et explias aspiscil esequi ius arum comnist emquam, amusda coria comnit et alis ea quae. Nam volupta ssincium quatemporum hilis plaboreium voluptatia ne pos dolenectur, commole stibus a quiam venis qui natur, que quis sunt.

10/12 Times New Roman

Vellacc usciusto temodi volupti sunt, utemolutempe ipsus, torit autem venda quiatem sam aceatem et explias aspiscil esequi ius arum comnist emquam, amusda coria comnit et alis ea quae. Nam volupta ssincium quatemporum hilis plaboreium voluptatia ne pos dolenectur, commole stibus a quiam venis qui natur, que quis sunt.

12/13 ITC New Baskerville Roman Std

Vellacc usciusto temodi volupti sunt, utemolutempe ipsus, torit autem venda quiatem sam aceatem et explias aspiscil esequi ius arum comnist emquam, amusda coria comnit et alis ea quae. Nam volupta ssincium quatemporum hilis plaboreium voluptatia ne pos dolenectur, commole stibus a quiam venis qui natur, que quis sunt.

12/13 Times New Roman

FOR TEXT, THESE DECISIONS ARE ABOUT:
- typeface (or font)
- typesize
- leading (or space between lines of type)
- measure (width of a line of type)
- depth (the number of lines on a page)
- margins

FOR ILLUSTRATIONS, IT IS ABOUT:
- position
- resolution
- use of color
- size
- shape
- quantity
- their relationship to each other and to the text

PERFECT BOUND

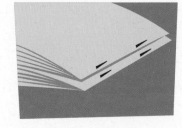

SADDLE STITCH BOUND

SEWN

Above: Publishers can choose from three different binding styles: perfect, or adhesive, binding in which the spine of the book block (the inside of the book) is attached directly to the cover with glue; saddle-stitched binding in which the book block and the cover are joined and held together by wire staples; and section sewn binding where the sections which make up the book block are joined together by thread before the book block is attached to the case or the cover.

While the designer is responsible for deciding how text and illustrations appear on the page, it is usually editorial and marketing who make the decisions that determine the visual and physical nature of the book, and ultimately the size of its environmental footprint.

THESE DECISIONS ARE ABOUT:
- the number of words in the book
- its format (dimensions)
- the use and number of illustrations
- the use of color
- whether illustrations are integrated or appear separately as plate sections
- whether it is a paperback or hardback
- binding style: sewn, perfect bound, or slotted

To understand how this works, it is helpful to work through this case study, which involves the biography of a famous person. (See opposite.)

So how possible is it for a company to be sustainable and be green at the same time? Sustainability in publishing is not necessarily about doing less, it is about doing it differently and smartly. Sustainability is about a company:

- striking and maintaining a balance between the costs and benefits of becoming green, and
- maximizing the trade-offs between the economic value of what they do and the economic value of reducing their environmental impact

With regulation increasingly affecting all aspects of the supply chain, and with better informed and more powerful consumer expectations, the time for not doing things in a green way is running out. In light of this, what could the company in the case study do to reduce the impacts of this particular product?

"While the designer is responsible for deciding how text and illustrations appear on the page, it is usually editorial and marketing who make the decisions that determine the visual and physical nature of the book, and ultimately the size of its environmental footprint."

CASE STUDY:

Marketing wants it to be the biography of the year, and the author has been contracted to write 200,000 words. The book is to be integrated (where illustrations and text appear together on the page throughout the book instead of in plate sections, which appear at intervals in the book). The format is 6.25 x 10 inches (Royal 8vo, portrait); it is to be printed in four colors throughout on a 100gsm coated paper and bound as a slotted hardback, with head and tail bands, and a four-color laminated jacket.

At approximately 480 words per page the text comes out at 417 pages, and together with eight pages of prelims and a six-page index it comes to 431 pages. To these should be added sixteen full-page, four-color illustrations; eighteen half-page, four-color illustrations, and twenty-four quarter-page, four-color illustrations. The total extent is now 462 pages, which will be printed as twenty-nine sixteen-page signatures (or sections) to produce a book of 464 pages.

From a commercial and a marketing perspective, this book is perfect. It is full of information (200,000 words and fifty-eight pictures), it's big (that is, long), and the production values are reasonably high. This is what the market wants, and the marketing department is going to make sure that this is what the market gets.

From a commercial perspective, it is difficult to argue against this: the company exists in a commercial environment and its primary purpose is to make money; to do anything less would be to threaten the company's sustainability, and for most companies sustainability is based on financial, not environmental, considerations.

But, as we shall see, from an environmental perspective its impacts are high throughout the supply chain.

Reducing the impact of products on the environment

As a word of caution: The options that follow all contribute to reducing product impact, but at the same time some of them may radically alter the nature of the product and how it was originally conceived. For these reasons not everything suggested here may be acceptable to editorial and marketing, even though they may be acceptable to the market. In spite of this, there is a wide range of options to choose from, which include:

ADJUSTING THE TYPOGRAPHIC PARAMETERS BY:

- **specifying a narrow typeface with a small x-height**
In the examples opposite, the lines of type are set in different fonts (typefaces), but are all in the same typesize (11 point). As you look down the list, you can see that they get longer and, although they are all set in the same size, some are larger than others, that is, they have a large x-height. For example, Perpetua (line 1) is the shortest and one of the smallest in appearance, while Verdana (line 7) is the longest and has the largest x-height.

Every word in this line

Verdana   Helvetica   Futura

1. The quick brown fox jumped over the lazy dog .... *(Perpetua)*

2. The quick brown fox jumped over the lazy dog .... *(Garamond)*

3. The quick brown fox jumped over the lazy dog .... *(Calibri)*

4. The quick brown fox jumped over the lazy dog .... *(Arial)*

5. The quick brown fox jumped over the lazy dog .... *(Georgia)*

6. The quick brown fox jumped over the lazy dog .... *(Palatino)*

7. **The quick brown fox jumped over the lazy dog** .... *(Interstate)*

8. The quick brown fox jumped over the lazy dog .... *(Verdana)*

The combination of narrowness and small x-height means that text set in Perpetua takes up roughly 25% less space than if set in Verdana. The extent of a book of 192 pages set in 11 point Verdana drops to 144 pages if set in 11 point Perpetua.

- **specifying a small typesize**
  For example, forty lines of text set in 11 point Calibri will occupy fifty-eight lines if set in 12pt Calibri—an increase of 40%.
- **widening the length of the text line (measure)**
- **increasing the number of lines on a page**

Where possible, it is certainly worth experimenting with these factors in different combinations to find out which works the best in producing text and pages that are economical in their use of space and easy and pleasurable to read.

You may find that you cannot adjust everything typographically at once, as the effects of doing so may not fit with the series design or brand. But being aware of what you can do is a necessary precondition for being able to achieve at least two of the three Ds: dematerialization and de-carbonization.

is Set in 32 POINT Type

AGaramond   Bodoni small caps   Times

Page size (mm) x Paper weight (gsm) x 1/2 the extent

$$\div$$

1,000,000
+ 20g for a paperback cover
+ 100g for a hardback cover

Above: Formula to calculate the weight of a book.

OTHER OPTIONS ARE TO:

• **Reduce the word count and the number (or size) of the illustrations**

For example, if the word count were reduced to 150,000 words, and the illustrations were reduced either in number or size to fit into the equivalent of twenty-four pages, the extent could be reduced to 352 pages–a reduction of just under 25%, with corresponding reductions in paper and energy consumption, and waste and pollution.

• **Reduce the paper weight**

The 464-page book printed on a 100gsm paper weighs 2.1 pounds. Printed on a 70gsm paper, the weight drops to 1.53 pounds. With an extent of 352 pages, and printed on a 70gsm paper, the weight drops to 1.21 pounds.

• **Reduce the paper thickness (bulk, volume, calliper)**

If the book, as currently planned, were printed on a Vol. 20 paper, it would bulk 1.83 inches between boards, though the weight would remain at 2.1 pounds. If printed on a Vol. 16 paper the bulk would drop to 1.46 inches between boards, and still weigh 2.1 pounds. With the extent reduced to 352 pages, and printed on a Vol. 20

70gsm paper, the book would bulk just under an inch between boards. If printed on a Vol. 16 70gsm paper, the bulk would go down to 0.77 of an inch (3/4"), though the weight would remain unchanged at 1.21 pounds. Thinner books take up less room in a container, on a pallet, and in the warehouse. However, it must be borne in mind that from a marketing point of view thick books represent value for money.

• **Print the illustrations as plate sections**

This option effectively turns the book from an integrated book, in which illustrations can appear close to the text that refers to them, to one where the illustrations appear in an eight- or sixteen-page section printed on different paper, between sections of text. Printing sections in this way has several effects:

• *on the book's design and organization*

In design and organizational terms, as the book is no longer integrated, the designer needs to treat text and illustrations separately. The reader now has to go to the relevant plate section to look for an illustration, rather than being able to refer to it at the point of reference.

Right: Dot gain occurs when printing on uncoated paper, and the halftone dot that carries the image swells as the ink is absorbed into the surface of the paper. The dots in the illustration on the right show signs of dot gain, as does the line of type above.

TYPE

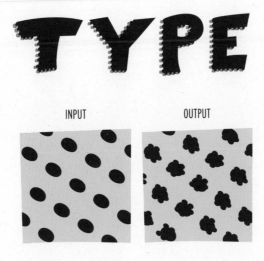

INPUT   OUTPUT

- *on how the book is printed and bound*
  In printing terms, the book now consists of text and illustrations, printed separately on different types of paper, only brought together during binding. In binding terms, care must be taken to ensure that the illustration sections are bound in at the correct point(s) in the book.

But, at the same time, it produces several benefits, both environmental and economic. The first of these is that printing the illustrations as three eight-page sections reduces ink usage as well as the cost of printing. The second is that it is now possible to use two kinds of paper: uncoated for the text and coated for the illustrations. For example, you could now use a 70gsm uncoated recycled paper for the text, and an 80 or 90gsm matte coated art paper for the illustrations, reducing paper costs and making the book lighter, which would not have been possible in an integrated book, where paper needs to be a uniform weight and finish throughout and capable of taking halftone printing. Using recycled uncoated stock for the greater part of the book reduces the book's environmental impact in terms of procurement—raw materials, energy, water, and

waste—as well as in terms of disposal: uncoated stock is easier to recycle than coated, and decomposes quicker than coated stock, should it happen to end up as landfill.

- **Print on uncoated paper**
  For monochrome, or single-color work involving halftone illustrations, one option is to print everything—illustrations and text—on uncoated paper. Doing this makes it possible to produce an integrated monochrome book on one type of stock, eliminates the need to print illustrations on separate coated paper and bind them in as separate sections, and reduces paper and binding costs and the use of raw materials, simultaneously. This option is quite popular, though the results can often be disappointing, as images lose their tonal qualities and become virtually indecipherable. This problem is caused by dot gain, which occurs when ink is transferred from the halftone dot to an uncoated paper and swells up (gains) as it is absorbed into the paper. The result is that highlights become mid-tones, mid-tones become shadows, and shadows become solids. This problem does not really occur with coated papers as the paper is coated with china clay to reduce absorbency as well as to make it

Above: Text and integrated illustrations printed in four colors.

bright white. The effect of dot gain can be eliminated by compensating for the paper surface when the images are being scanned. By doing this, a halftone dot, even though it still gains when it is absorbed, gains so that it ends up at the required size—and highlights remain highlights, mid-tones remain mid-tones, and so on. This can also be done with color—after all, newspapers and magazines do it successfully, and it is now quite a common feature of books, especially picture books. The only sacrifice is in the brightness and whiteness that come with using coated stock, but this is a small price to pay if the result is a reduced environmental footprint.

• Reduce the amount of color

This is done by grouping color illustrations and black-and-white illustrations so that they appear as alternate spreads. This allows the printer to print the sheet four-colour on one side and a single color on the reverse (called 4/1 printing, and pronounced four-back-one), as opposed to printing 4/4 (four-back-four), where the sheet is four-color both sides. This requires quite a lot of organization to make sure that everyone concerned—the author, designer, and illustrator—are aware of which illustrations are going to be in color, and where they appear, and which are going to be black and white.

Above: The same text and illustrations printed in two colors.

• Have pictures that do not bleed

Pictures that bleed (known as bleeds) go up to and continue over the edge of the page, so that when they are trimmed they end exactly at the edge of the page. To achieve this, it is necessary to print on bigger paper than would be needed for the same job if it did not have bleeds. The extra needed is generally about 1/4 inch along each edge for trimming and to make it possible for the grippers to pull the sheet through the printing press if it is sheet-fed. For a Demy 8vo book (5.63 x 8.75 inches, portrait), the extra amount of paper required is 7%, which is literally trimmed off and disposed of as waste. The worst case scenario is to have one or two bleeds in each section, which means that the entire book has to be printed on outsize sheets. It is easier either to have no bleeds at all, or to have as many as are needed. It is difficult to print a product that partially bleeds.

• Avoid solid colors

A solid color is one that has no gradations—for example, highlights—and it usually covers a large part of the page. The reason for avoiding solids, where possible, is to reduce ink, and hence raw materials usage, and to make it easier to recycle. If the sheet goes into landfill instead, there is an awful lot of ink, vegetable- or mineral-based, that will eventually find its way into the ecosystem.

Above: The same text and illustrations printed in a single color.

- **Use certified recycled paper wherever possible**
  There is an enormous range of recycled papers available to choose from that will more than adequately satisfy your production requirements—from newsprint through to quality art paper. As mentioned in Chapter 5, you should make certain that it is properly certified to meet the standards set by organizations like FSC or PEFC. Within the range of recycled papers available, experiment and try to use lower grades wherever appropriate, bearing in mind that the higher the grade of paper the more raw materials have been used to make it.

- **Bind the book as a paperback**
  Unless there are marketing reasons for binding the book as a hardback, you should think about binding your books as paperbacks. They are lighter than hardbacks: a paperback cover weighs about .7 ounces and a hardback case about 3.5 ounces, added to which there is a jacket, which has to be wrapped around the bound book and uses raw materials. In short, paperback books cost less to produce than hardbacks both financially and environmentally.

Right: Soft proofing makes it possible to view and correct proofs on screen rather than on paper. It is inexpensive and effective, and allows for instant feedback when problems arise.

• Stick to standard formats

Again, unless there are strong marketing imperatives for using a non-standard format, it is more environmentally effective to stick to standard formats. Book printing presses are built to maximize the use of standard formats: Royal format books are most economically printed on presses that take Royal size sheets. Of course, it is possible to print Demy books on Royal size presses, but it is wasteful of paper, plates, and energy. This is worth bearing in mind when thinking about that new title. If it *has* to be 7 x 7 inches, then there is no contest. But if this is not the case, then spare a thought for the environmental consequences and go for something standard. (For a list of standard formats, please refer to the appendix.)

- **Use soft proofing**

Digital proofs can either be printed onto paper, in which case they are known as hard proofs, or they can be viewed and annotated online for correction by the supplier. This is known as soft proofing. In a hard proofing workflow the proofs are printed, delivered to the author, annotated, and returned to the supplier who applies the corrections; and, if necessary, the cycle is repeated until copy is judged to be ready for press. Hard proofing is labor intensive, comparatively slow, uses paper and ink/toner, and proofs may have to travel long distances between the publisher and author and back again. In soft proofing, the proofs are uploaded to an FTP site or a central proofing system where they can be accessed online and viewed by whomever needs to see them, corrections are annotated, and applied by the supplier. The process is quick and does not require paper or ink/toner.

- **Use digital printing**

By using digital printing it is now possible for publishers to exercise tighter control over the number of copies to print at any one time and to order the exact quantity of stock they need whenever it is needed. Before this was possible, publishers had to make an educated guess at how many copies to print. Sometimes this would be just right, while at others it would be hopelessly inaccurate, with the result that there were either not enough books to meet demand, or too many, which, as discussed in

Chapter 4, leads to the problem of returns. Digital printing now allows a publisher to print little and often, just in time (JIT), or single copies as and when needed. This is possible because digital technology lends itself to single-copy production in a way that lithographic printing (also known as litho or offset) does not and cannot. Although offset printing has made remarkable progress in narrowing the gap and was, until recently, economically viable for 600 copies and above, this is no longer the case: digital printing is now economical for over 5,000 copies, brought about by the existence of digital presses that can print and bind a single copy of a 1,000-page plus book in under a minute.

When compared with offset printing, the immediate environmental benefits of using digital printing are reductions in waste and emissions from print processes and raw materials, water, and energy consumption. Other environmental benefits are reductions in the number of unsold copies making their way into landfill, because of greater flexibility and accuracy in deciding on print runs; and reduction in transport costs and book miles, as fewer copies are moved around the globe, because stock can be printed closer to its point of end use through distributed printing (see p. 146).

Digital printing goes a long way towards reducing some of the risks associated with publishing a book, and one way that it does this is by making the manufacturing process more flexible and quicker at responding to demand. Recent

Above: A digital press that can produce a finished copy of a book in one operation, avoiding plate-making and all the other operations—like binding and finishing—that would otherwise be involved in producing a book by non-digital means.

developments mean that quality, especially in four-color work, is now comparable with the quality achieved with litho. It is true that digital printing has a measurable environmental footprint, but what industrial process hasn't? And book for book it is a great deal smaller than with litho. These options are presented in much the same way as an à la carte menu from which you can make choices, while reserving others for later on. To implement them all at once might be a step too far, too quick, but to ignore them would be irresponsible.

In *Book Production* (2012), I say: "Becoming environmentally aware, and doing environmentally friendly things, is a gradual process, and it doesn't happen overnight. You don't have to do everything at once, and not all your products necessarily lend themselves to being printed on recycled paper, in small quantities, in far flung corners of the globe. Starting with a few titles, and going from there is a good start; and, at least, it is a step in the right direction." [Bullock, p. 159.]

Nevertheless, as the Green Press Initiative points out: "Publishers have the opportunity, through the products they choose and those they reject, to serve as environmental stewards in improving the production practices of the entire book publishing sector."

A start needs to be made somewhere, and goals and deadlines need to be set. So for example, when it comes to paper, the Green Press Initiative suggests minimum benchmarks for publishers to start including post-consumer recycled fiber in their products, which cover a five-year period, beginning in year one with selected titles and progressing by year five to the point at which paper usage should consist of at least 30% recycled FSC-certified fiber.

OPTIONS FOR NEWSPAPERS AND MAGAZINES

Before closing the chapter, it is important to look at how and where newspapers and magazines fit into all this.

Newspapers and magazines, as well as being physically different from books, are unlike books in other ways too: for example, books are essentially one-off, customized micro-products, each of which is a unique combination of content, design, and function, which makes them relatively easy to adapt on a book-by-book basis to make them greener and more environmentally friendly.

Newspapers and magazines, by contrast, once set up, tend to stay that way, changing as little as possible. So introducing some of the more radical options suggested above is not really an option.

The one option that is open to them is to produce as much of their output as possible on recycled fiber, and many newspapers already do this, which, given the enormous amounts of paper they consume each day, is just as well. The other option is to move away altogether from print-on-paper products in favor of digital ones; and, again, several newspapers, closely followed by the magazine sector, have already started doing this.

In this chapter we have looked at incorporating green practices into the early decision-making process of how publications are designed and developed, with a focus on striking a balance between producing a publication that meets the customer's needs and has a reduced environmental impact.

"Becoming environmentally aware, and doing environmentally friendly things, is a gradual process, and it doesn't happen overnight. You don't have to do everything at once, and not all your products necessarily lend themselves to being printed on recycled paper, in small quantities, in far flung corners of the globe. Starting with a few titles, and going from there is a good start; and, at least, it is a step in the right direction." (BULLOCK, P. 159.)

7

Eco-friendly printing & binding

We have now reached the point in the supply chain when we need to find suppliers–printers and binders–to turn everything that we have been working on into a finished product. So far we have done all we can to keep our product green by integrating and applying green principles and practices into the way we have conceived and designed it. Now we want to make sure that it stays green by choosing suppliers who are able to demonstrate that they can work with us to print, bind, and deliver stock that not only is green and has a reduced environmental impact, but was produced using green technologies and green raw materials, meets green regulatory criteria and standards, and satisfies consumer expectations.

Generally speaking, the supplier that is most likely to be able to do all this is one that is green. But, as mentioned in Chapter 4, there are degrees of greenness ranging from dark to light–and the darker the green, the better for the environment.

What makes a supplier green?

So what does make a supplier, and a printer in particular, green, exactly?

A printer is green because they understand and are able to manage their impact on the environment through an accredited environmental management system (EMS). They are also green because their claims about their products and their environmental performance meet an accredited standard.

The first thing you have to do is establish for yourself the extent to which they do this, and how successfully and effectively they do this when measured against their environmental performance indicators. This means asking the supplier to provide you with information about their environmental credentials, their environmental policies, and their environmental performance.

The second thing to do is audit these against the green printer checklist shown opposite. The more ticks a supplier earns the greener they are.

Environmental management systems (EMS) and the regulatory aspects of being green have already been dealt with elsewhere in this book, and it is easy to see why they reappear as items in a checklist designed to audit supplier greenness.

We are now going to explore the technological criteria so that we can understand the differences between the various manufacturing technologies offered by suppliers, the kinds of impact these technologies have on the environment, and why one technology may be greener than another. Doing this makes it easier for you to make informed decisions about which technologies to use for which kinds of work, and easier to reduce your own environmental impact.

	YES:	NO:
A GREEN PRINTER:		
1. has an EMS certified to a recognised national or international standard (ISO 14001, EMAS (Eco Management and Audit Scheme), BS 8555 (Guide to the phased implementation of an environmental management system including the use of environmental performance evaluation), Green Mark)		
2. can make claims about their products and environmental performance that meet ISO 14021 and ISO 14021 (Environmental labels and declarations: Self-declared environmental claims) or Green Claims Code standards		
3. has a carbon neutral scheme to offset emissions		
4. uses computer to plate (CTP) or processless plates		
5. has eliminated or reduced use of isopropyl alcohol (IPA) in their printing		
6. uses waterless printing		
7. uses an automatic wash-up system for their presses		
8. offers digital printing for small jobs		
9. uses vegetable-based inks/ultraviolet (UV) inks		
10. uses paper from certified sources (FSC/PEFC)		
11. has Chain of Custody (CoC) certification for FSC and PEFC papers		
12. can print FSC, PEFC, Möbius strip, and Recycle logos on your products		
13. has a waste reduction policy		
14. has a waste recycling policy		
15. has an energy policy		
16. has calculated their direct carbon footprint, and the carbon footprint of their products		
17. has a control of substances harmful to health (COSHH) policy		
18. has an air-pollution reduction policy		
19. has water-use, pollution, and waste reduction policies		
20. has a green transport policy		
21. uses an up-to-date set of environmental performance indicators (EPI)		
In addition, a green printer will be able to supply information on:		
· the provenance and production of the papers they use		
· the CO_2 emissions associated with your job		
· comparable prices for virgin and recycled papers		
· the bio-degradability and recyclability of the materials they use, for example, paper, boards, ink, plates, plastics, laminates, and packing		
· sustainability issues		

Printing & binding technologies

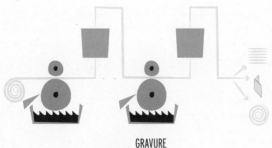

GRAVURE

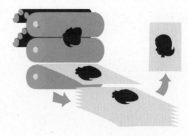

SHEET-FED LITHO

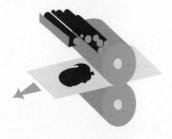

LETTERPRESS

Above Left and Left: Gravure and letterpress printing both have their uses—gravure for high-quality color work, particularly in magazines, and letterpress for fine editions.

There are several printing technologies to choose from, including offset lithography (litho for short), digital, letterpress, and gravure.

Although each one has its uses—letterpress for fine printing and limited editions, and gravure for magazines—litho and digital are the predominant technologies for book and magazine work, and we shall concentrate on these.

Litho and digital printing both produce books as the end product, but how they do this is radically different, and with eco footprints that differ markedly in size.

DIGITAL PRINTING

A digital press works like a desktop printer, transferring the image to paper, a sheet at a time, either using toner if it is laser printer, or a water-based ink if inkjet. Like the desktop printer, the digital press is ready to print whatever you want whenever you need it—there is no set-up time or make ready, and there are no plates. The latest digital presses, if set up with an in-line binding facility, are capable of producing, in less than a minute, a single copy of a single-color, perfect bound, 1000+ page book, complete

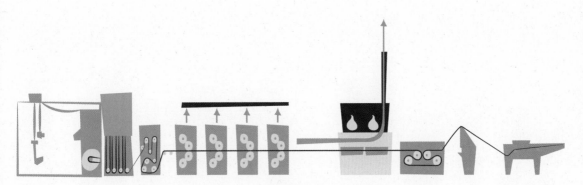

WEB-FED LITHO

with a four-color cover, and this makes it the technology for single copy (print on demand or POD) and short-run printing. Until very recently the optimum number for digital short-run printing was between 350 and 400 copies before it was more economical to change to litho. However, recent developments have raised this to about 5,000 copies, and counting. The speed and falling costs of digital printing alone make it a serious contender with litho as the first choice for publishers to print their books —both front and back list—and this is already happening. But as we shall see, there are environmental benefits too.

LITHO PRINTING

Where digital printing produces a single bound copy of a book at a time, litho printing mass produces the components of a book—printed sections—that have to be assembled and joined together to make a book during binding.

Litho printing can be done in one of two ways. The first is sheet-fed printing and the other is web-fed. In sheet-fed printing flat sheets of paper pass through the press, are printed, and delivered at the other end as flat sheets. In web-fed printing, paper passes through the press as a continuous strip (web) from a reel. Once printed, and while still on the press, the paper is cut into sheets that are delivered as folded sections.

In both methods, the image is first transferred to a printing plate from where it is transferred to the printing surface using ink. Pages are printed in groups of four, eight, or sixteen at a time on each side of the paper (sheet). A sheet printed with a group of four pages on each side (known as printed to view) contains eight pages; printed eight pages to view it contains sixteen pages; and so on. A 320-page book, printed sixteen pages to view, uses ten sheets of paper. Pages one to thirty-two appear on sheet one, followed by pages thirty-three to sixty-four on sheet two, pages sixty-five to ninety-six on sheet three, and so on right through to page 320.

Above and Opposite: Sheet-fed and web-fed lithographic printing is still widely used for printing books, especially for long runs. Their dominance, however, is being eroded by digital printing, which is fast, versatile, and cheap.

Printing the book sixteen pages to view uses twenty printing plates—one for each side of the sheet and each plate needs to be:

- processed before it can be used, which involves chemicals and energy
- fitted to and removed from the press before and after use, which takes time
- made ready so that it prints properly, which also takes time
- disposed of—printing plates are usually metal (aluminum) which can be recycled

Only when printing has been completed is it possible to start the process of converting the output into books, which starts with folding the ten sheets into sections. (With web-fed printing this is unnecessary, as folded sections were delivered as the output.) The folded sections are then gathered in the correct sequence to form the book block, which is either glued or sewn together, before it is finished as a softcover or hardcover, and packed and dispatched to the publisher's warehouse.

DIGITAL OR LITHO: MAKING THE CHOICE

Both litho and digital printing are industrial processes that, by their nature, use raw materials and create waste and pollution, though not to the same degree, as we shall see. That being the case, you would expect the choice to be a no-brainer, and go for the greener technology every time. However, both technologies offer features and benefits that

the other cannot, and they both have their uses, which allow them to exist side by side. Though for how much longer this will continue remains to be seen. To understand why, we should look at the business and the environmental case for each technology in turn. The economic and environmental case for digital printing is that:

- there is virtually no set-up time; this reduces cost, time, the use of raw materials, and waste

- it is possible to print single copies on demand, very quickly; this reduces the risk associated with printing large quantities of books that may not sell, with the added benefits of reducing:
  - financial risk
  - inventory
  - the unnecessary use (and waste) of raw materials
  - unnecessary emissions
  - the quantity of unsold copies (returns) being returned by booksellers to publishers, which, unless resold, may end up in landfill
  - unnecessary transportation and book miles
- it is possible to print where demand and the markets are by using distributed printing. This speeds up order fulfillment at the same time as it reduces book miles. For example, a print-on-demand request from New Zealand for a book published in the U.S., if printed digitally in the U.S. and sent by air to Auckland, clocks up 8,820 miles, which would not be the case if printed digitally in Auckland. Digital printing has its limitations, which have

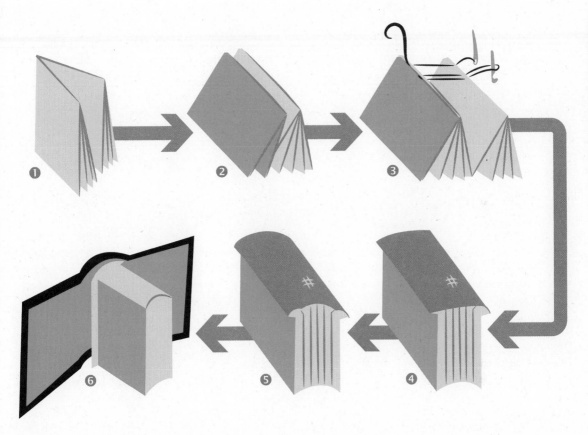

mainly to do with cost and quality, though costs are coming down and quality is improving.

The main planks of the business and environmental case for litho printing are that:
• it is capable of producing long print runs–5,000 copies upwards–cheaply and relatively quickly. For example, it would not have been practicable to have produced the record-breaking print run of 12,000,000 copies of *Harry Potter and the Deathly Hallows* other than by litho
• it is capable of producing high quality work
However, litho printing has its limitations:

Above: The stages a printed section passes through on its journey to becoming a hardback book.

① Folded section

② Endpapering

③ Sewing

④ Lining

⑤ Rounding and backing

⑥ Casing in

- turnaround is slow—this makes it unsuitable for on-demand and really short-run printing
- costs are front-loaded—this makes it uneconomical for short-run printing
- it is labor-and equipment-intensive, which increases costs, but is ideal for dealing with long runs
- its use of raw materials and energy is higher than in digital printing. For example:
  - printing plates are not used in digital printing
  - paper: has higher spoilage rates (see below)
  - ink
  - solvents
  - water
- it creates more waste than digital printing in terms of:
  - paper
  - ink
  - solvents
- it releases more emissions—VOCs—into the atmosphere than digital printing

WATERLESS LITHO

If you do decide to go litho, there is one way in which you can help reduce the impact of this choice, and that is by choosing a printer who offers waterless litho.

Waterless litho has the same business limitations as ordinary litho, but its green credentials are higher because it:

- does not use an isopropyl alcohol-based solution to damp the plates, so there is no release of VOCs
- uses waterless and vegetable-based inks, so there is no release of VOCs
- eliminates the need for water-ink balance, which speeds up make-ready and turnaround, and reduces waste

On balance, digital is the greener option, and one that publishers are increasingly turning to for their printing requirements. However, it is not always the case that digital is preferable to litho, and before making your final decision it is certainly worth considering and, if necessary, discussing with the supplier the pros and cons of each technology to see how well it matches up with what you are trying to achieve, as well as how fit it is for your purpose, how long it will take, and how much you are prepared to pay.

"It is important to keep sight of the fact that becoming green is a process, and that it takes time."

BINDING

Binding, as already described, is either an equipment- and labor-intensive process if it is dealing with output from a litho press, or it can be done as part of the digital printing process. How you bind is governed by how you print.

Books are bound either by sewing or gluing, and the end product is either paperback or hardback. Both methods use glue.

If you are binding litho output, the impact is going to be high because of the nature of the processes involved, and your ability to limit your environmental impact is pretty much limited to the type of glue (or adhesive) you use. There is a vast range of glues to choose from, starting with animal glues (which are certainly environmentally friendly because they are bio-degradable, but may not be acceptable to everyone) and starch-based glues through:

• water-based emulsions (known as cold-melts) based on polyvinyl acetate (PVAc) and polyvinyl acetate ethylene (VAE) copolymers
• hot-melts, based on ethylene vinyl acetate (EVA) copolymers
• reactive hot-melts (also known as warm-melts), based on solid pre-polymer polyurethanes or PURs

Both the hot-melt and the cold-melt glues serve different purposes and work better with one type of paper and binding style than another. The water-based cold-melts are more environmentally friendly than the hot-melts as they use less energy and release fewer emissions. Binders often use them in combination, especially in paperback binding.

PUR glues are becoming increasingly popular with binders not only because of their binding properties, but because they are environmentally friendly and have a lower melting point than EVA glues. They also degrade relatively easily, which is useful when it comes to recycling.

It is important to keep sight of the fact that becoming green is a process, and that it takes time. The Green Press Initiative suggests a five-year program for converting paper use from virgin to recycled, starting with a selection of titles and gradually building this up. The same approach can be applied to printing and binding. In the end not everything is a candidate for green treatment, and deciding on whether to use digital or litho for printing your products must depend on a number of factors in addition to the purely environmental ones.

Where should you print?

Globalization and digital technology mean that it is now possible, and very easy, to get your printing and binding done anywhere in the world. It also means that you can move your work to wherever you are going to get the best prices. For the moment, at least, and certainly for color work, this tends to be the Far East, India, and the Gulf, as a quick look at the title verso of a book will reveal.

From a financial point of view, printing where it is cheapest, even if it is far away, is a sound thing to do. From an environmental point of view, this is not necessarily so: the main problem being the distance the finished products have to travel back to the publisher's warehouse so that they can be distributed.

This is not an argument for buying print on your doorstep—though this is happening with the just-in-time and print-on-demand business models. It is an argument for rationalizing your print buying and doing things smartly by printing close to where the main markets are and using distributed printing; and by doing so reducing the number of miles your books need to travel before they reach their final destination. The case study on p. 140 shows how this can work.

HONG KONG TO NEW YORK 12,691 MILES

Right: Sample distances from printers to publishers.

OXFORD TO SHENZHEN 6,012 MILES

DUBAI TO NEW YORK 9,159 MILES
BOMBAY TO NEW YORK 9,483 MILES

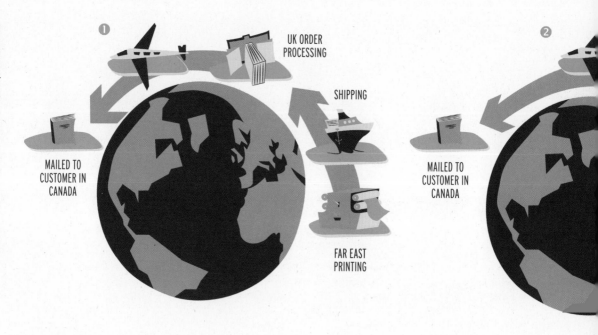

① UK ORDER PROCESSING

SHIPPING

MAILED TO CUSTOMER IN CANADA

②

MAILED TO CUSTOMER IN CANADA

FAR EAST PRINTING

CASE STUDY:

A UK-based medical publisher had successfully published a book on palliative care for the UK market and wanted to publish a version written specifically for the Canadian market. The publisher was faced with two immediate options:

• print books in the UK, process orders in the UK, and mail books to customers in Canada

• print books in the Far East, bring them back to the UK, process orders in the UK, and mail books to customers in Canada

A third option was to find a printer in Canada able to print the books, process orders, and post books to customers in Canada, maintain records of sales and payments, and remit money to the UK publisher. The production department searched the internet, and found a printer in Canada, just outside Toronto, who was prepared to do this. Everything was set up, and the arrangement continues to work successfully to this day.

By doing this the publisher was able to save thousands of individual book miles (the distance between the printer in Toronto and the publisher is 3,500 miles) and the environmental degradation they would have caused. They were also able to save the postage costs of sending 1,750 copies across the Atlantic to individual customers all over Canada. They saved the book miles that would have been clocked up by printing the book in the Far East and bringing copies back to the UK, and finally, they were able to make significant reductions in the time taken between the customer ordering their copy and receiving it. This project ended up being a win-win situation for the publisher and, just as importantly, for the environment.

Not every job can be as clear cut as this one was, however. Nevertheless, if you are not doing this already, it is certainly worth considering for some of your work. Although their business model is not strictly comparable with book publishing, journal and magazine publishers are already using distributed printing extensively. A trail has already been blazed and the concept is no longer new; all it needs now is for book publishers to follow.

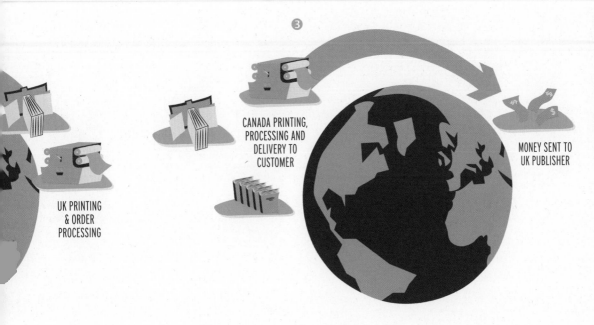

UK PRINTING
& ORDER
PROCESSING

CANADA PRINTING,
PROCESSING AND
DELIVERY TO
CUSTOMER

MONEY SENT TO
UK PUBLISHER

"From a financial point of view,
printing where it is cheapest, even
if it is far away, is a sound thing to do.
From an environmental point of view,
this is not necessarily so."

Once you start to build up a range of green suppliers, all working towards the same target–the reduction of your, and their, environmental footprint–things become easier. You gain critical mass; and inputs, raw materials, outputs, design, manufacturing, and distribution gradually, but surely, become greener and environmentally friendlier.

Above: In the case study, the publisher has three options. The first, and most expensive environmentally, is to print in the Far East, bring stock back to the UK, and process and fulfil orders from there. The next is to print in the UK, process and fulfil orders from there. The third, and the least expensive both commercially and environmentally, is to print stock in Canada, process orders in the UK, and fulfil them from Canada.

8

Cutting back on transport & travel

By the time we reach this chapter we should know quite a lot about what we can do to make our products and our supply chain as green as possible. By now we should understand that it is possible to choose raw materials and design products that do not have as great an impact on the environment as those that are not sourced or designed on green principles. For our manufacturing, we are now able to define what makes a supplier green and can identify suppliers who are as green as they claim they are and we would like them to be, and who are able to deliver products that are commercially viable while being green at the same time.

Printing overseas

Any business that wants to continue in existence needs to keep an eye on the bottom line, and will, understandably, do all it can to ensure that it takes advantage of whatever options there are to cut costs—and one of these is the cost of production. In a globalized economy, and with the help of the internet, it is possible to move production around the world in search of the cheapest prices. Today it may be the Far East; tomorrow it could be South America.

For something like a book, because the equipment, skills, raw materials, and transport needed to produce and deliver high quality products are relatively easy to find, it is quite simple to shift production quickly from one country to another, which cannot be said to be the case when it comes to producing computers or cars.

Sending work to printers around the world has become even simpler since the arrival of the portable document format file (PDF) which, with the internet, makes it possible to send complex information—text as well as graphics—anywhere in the world in less time than it takes to read or write this sentence.

Sending work to distant places to be printed and bound is not inherently a problem—indeed, it makes good commercial sense and, if properly thought out, can make good environmental sense as well. The problem arises, commercially as well as environmentally, when publishers bring their books back to base, only to send them out on another journey to their point of use.

For example, by printing and binding their books in India, a UK publisher can reduce their manufacturing costs by 25-30%, which is a powerful incentive at the best of times, and even more so when times are hard. The publisher asks for the books to be transported from Chennai to their warehouse in Oxford, 7,961 miles and three to four weeks away by sea.

Books have to be packed, containerized, and put onto a ship, which costs money, takes time and effort, and uses raw materials in the form of:

- paper, cardboard, and film wrap for secondary packaging
- metal and wood for the containers and pallets needed for transport packaging
- energy
- water

1,550 MILES

980 MILES

1,050 MILES

7,900 MILES

12,600 MILES

13,242 MILES BY SEA

This produces waste and pollution in the process.

Once the stock is in the warehouse, order processing begins. For a new title these orders will have built up over a period prior to publication, a result of promotion and advance subscriptions from wholesale, retail, and individual customers. As each order is processed, books that have already traveled nearly 8,000 miles to get to Oxford now set off on their travels again.

For the UK and Europe the distances from Oxford are not that great—Moscow, 1,550 miles; Naples, 1,050 miles; and Lisbon, 980 miles—a rough average of 1,200 miles. But what about orders for India, China, and Australia where the distances involved are much greater? Delhi is 7,900 miles; Beijing, 12,600 miles; and Sydney, nearly

13,250 miles—a rough average of 11,250 miles, or a ratio of nearly 1:9. In terms of the environment, there is a significant price tag to pay for those book miles, and there is an economic one as well.

Above: Distances within Europe average about 1,200 miles between the furthest points of Moscow, Naples, and Lisbon. These distances are dwarfed when you take the distances between, say, China, India, and Australia, which average 11,250 miles—a ratio of nearly 1:9.

Distributed printing & distributed distribution

Interestingly, the same technology that makes it possible to source competitive printing prices around the world is also there to provide a solution to the problem. It is not the technology that needs to change, it is the mindset and the behavior that goes with it. There are signs that this is beginning to happen as publishers reduce the sizes of their initial print runs and print little and often, or just in time (JIT). (This has already been covered in Chapter 7.)

In the JIT model, to take an example, the publisher will produce an initial print run of a new title of, say, 500-750 copies to meet orders that have built up, with an extra 10% for the unexpected. This model affords the publisher greater control over their print numbers and inventory and an accompanying reduction in:

• the risk and impact of investing in stock that might not sell
• financial exposure
• the cost of holding inventory
• the cost of warehousing

The success of this model depends on rapid response and quick delivery, which means that printing is more likely to be digital than litho, and producing stock locally is more likely than overseas, because of the time it takes to bring stock back to base.

If stock is to be produced overseas, there are two production models worth considering, as they both contribute towards reducing the impact of the distance stock has to travel between the point of production and the point of use. The first is the distributed printing model, in which stock is printed close to the major markets and distributed from there. For example, a U.S.-based publisher is about to publish a book that has a market in the U.S., Australia, New Zealand, and the UK. In this model, a contract is set up with a printer in each country to produce stock from a PDF supplied by the publisher, and to distribute stock locally in response to orders received directly from the publisher. Financial arrangements are set up so that the printer invoices the customer locally and remits the proceeds to the publisher in return for an agreed transaction fee. Clearly, this model requires organization and management, but there are savings to be made, both financially and environmentally: response is quick, and books miles are significantly reduced.

The second is the distributed distribution model, in which stock is printed overseas using a single printer, and part of it is kept by the printer or moved to a local wholesaler (or similar organization) for local and regional distribution. In this model, for example, a UK-based publisher prints 7,500 copies of a title in Hong Kong: 5,000 copies are brought back to the UK for distribution

DISTRIBUTED PRINTING

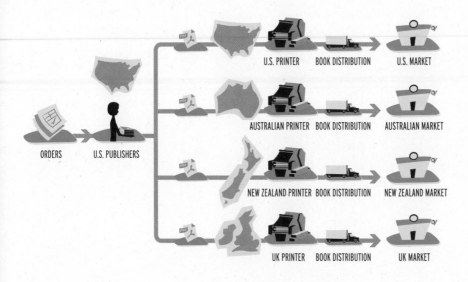

DISTRIBUTED DISTRIBUTION

UK PUBLISHER

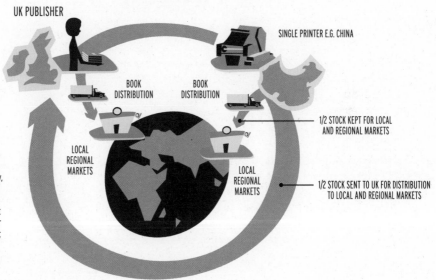

SINGLE PRINTER E.G. CHINA

1/2 STOCK KEPT FOR LOCAL
AND REGIONAL MARKETS

1/2 STOCK SENT TO UK FOR DISTRIBUTION
TO LOCAL AND REGIONAL MARKETS

Above: In the distributed printing model, stock is printed and distributed as close as possible to where the major markets are, something which has been made possible by the powerful technology available. Producing and distributing stock this way saves time, saves money, and saves the environment. In the distributed distribution model, stock is printed in one place. Part of the stock is kept by the printer for distribution within the region; and part of the stock is returned to the publisher so that they can service their region. This model, too, produces financial and environmental savings.

in the UK, Europe, and the USA. The remaining 2,500 copies are kept by the printer for distribution in China, India, Australia, and New Zealand, again for an agreed fee per copy. As with the distributed printing model, this needs some organization and management, but the advantages and savings outweigh the initial costs.

Both models are based on the smart use of digital information, communication, and manufacturing technology, and there is no doubt that either of these models is a preferable and viable alternative to bringing everything back to base and unpacking it, only to repack it and send it out again.

Cutting back on travel

We have been looking at ways of cutting back the distance our products travel; now it is time to look at how we can cut back on our own travel.

It has never been easier than now to travel long distances in the U.S. and UK—that is, provided you use private transport, and air travel has reduced the distance between cities to a matter of hours, to the point that we are more used to describing journeys in hours or days and not in miles. Distance is virtually meaningless.

Air travel is a major contributor of greenhouse gases and other pollutants to the atmosphere. George Monbiot, in an article in *The Guardian* newspaper in September 2006, wrote:

"Aviation has been growing faster than any other source of greenhouse gases. Between 1990 and 2004, the number of people using airports in the UK rose by 120%, and the energy the planes consumed increased by 79%. Their carbon dioxide emissions almost doubled in that period—from 20.1 to 39.5 megatonnes [22.1 to 43.5 megatons], or 5.5% of all the emissions this country produces. Unless something is done to stop this growth, flying will soon overwhelm all the cuts we manage to make elsewhere. But the measures the government proposes are useless."

In production we travel for a variety of reasons:

- to visit a printer's factory
- to inspect sheets coming off the printing press and make sure that color values match the proofs
- to attend a conference
- to negotiate prices and sign an agreement

Before you undertake that next journey, ask yourself if there are other ways of communicating with the printer than face-to-face onsite.

As discussed earlier in this book, it is now possible with the kind of communications technology that is available to check proofs and printed output in real time on-screen, or by email. It is possible to talk to printers by phone or over the internet using webcam technology and to find out about what they are able to offer by visiting their websites. As an example, a recent medical textbook was written and edited in Oxford and Nottingham, typeset in India, and printed in Canada. Proofing was done entirely online, and editors annotated PDF files of each chapter for correction by the typesetter. The final PDF file was sent to Canada. The Canadian printer was found using the internet, and the entire job—estimating, scheduling, and issuing production orders—was managed by email. The printer insisted on the PDF file being checked by the publisher before it was printed. This could have been done by sending a hard copy proof to the publisher by courier, which would have added at least ten days to the schedule;

"Air travel has reduced the distance between cities to a matter of hours, to the point that we are more used to describing journeys in hours or days and not in miles. Distance is virtually meaningless."

instead, the printer set up an FTP site, which allowed the publisher to check the proofs on-screen. It was agreed that the printer would accept responsibility for ensuring that printed and bound outputs met the required standard, which eliminated the need to see printed sheets (running sheets) from the main run or advance copies—a practice that is being adopted increasingly by publishers, as more printers adopt ISO standards for quality management.

For large gatherings of people, for example a conference or a negotiating meeting, there is teleconferencing and other media like the access grid technology that make long distance communication possible and pleasant, and in doing so they eliminate the need to travel and help us reduce our carbon footprint and its impact on our world.

We have now come to the end of the section in the book that looks at the practical and organizational aspects of going green. In the next chapter we look at the environmental issues.

9

Beyond the door

What are the environmental issues at stake once a book has been produced, and what are a publisher's responsibilities after production? An environmental Lifecycle Analysis incorporates the end life of a product; in publishing, this will include the returned products and their disposal. In this chapter we will talk about the issues created by "returns" on the different parts of the supply chain and about solutions for all parties. We will talk about several disposal options, the benefits of each, and how changes in disposal could lower the overall environmental impact of publishing and create a new form of product cycle. We will also look at the relationship between the publisher and the customer on environmental issues from production to disposal, and how this relationship may evolve.

Returns

That booksellers are able to return to publishers unsold books or magazines benefits the publishing industry; it allows for greater diversity of material sold by booksellers. Returns encourage diversity in book publishing by allowing the bookseller a safety net when they choose untried authors or unusual subject material, and therefore returns serve a useful and important purpose. They do however present a heavy environmental and economic impact. If a title is returned and can't be sold or used elsewhere, then this is a case of over-production and, as a consequence, wasted materials and energy: trees have been cut down unnecessarily with all the attendant ecosystem impacts on biodiversity; water has been used and its pollutant load potentially added to; and CO_2 has been created in the manufacturing and transportation of the books. The environmental and economic cost of returns is recognized in the U.S. and UK, and the issue is being addressed by organizations such as BIEC, the Book Industry Communication, and the BA/PA Environmental Action Group. The returns rate has fallen (in the UK the returns rate fell from 15.6% in 2007 to 14.2% in 2011,[21] though this statistic doesn't show the level of reuse within the system, an idea that we will explore shortly). However, while the issue has been addressed, the industry feels there is still more that can be done and that the returns rate could be lowered further, as the impacts remain substantial for business and the environment.

The return of books has business impacts for booksellers and publishers alike. For the booksellers, they have cash tied up in stocks, which is problematic in a highly competitive marketplace with potentially low price points and discounts already in effect. And resources tied up in stocks have a potentially irrecoverable value. Weighed against this is the need to get books into the marketplace and to the customer at great speed. The titles need either to be in the shops with the ability to replenish stocks quickly, or, if they are purchased online, to be despatched within seventy-two hours or sent to a bookshop for collection. This is what the customer has come to expect, and fulfilling this is felt to be necessary for customer retention. There is a second financial impact for booksellers. The cost of processing the returned stock is high: stock needs to be reviewed and processed by individual bookshops and through the retailer warehouses, incurring costs in both employee time and system management. Despite the cost of holding stock, booksellers feel the need to have stock in their shops and warehouses to meet their customers' requirements.

Returns create multiple costs for the publisher. The publisher pays for the cost of production, storage, transportation, processing (the initial order and the return), and pulping. These costs, incurred once they have left the publisher's gate, can be managed in part through print decisions. In making print decisions, the unit cost of a book in production will play an important part. It is possible, however, that this unit cost will not factor in warehousing, transportation, and returns, and within a company this overall picture may not be visible as these

Below: Unsold books are returned to the publisher

points of the supply chain will be handled by different roles in the organization, and so these elements are hard to incorporate into a print decision. If these factors could be brought into individual print decisions, would it change the way we think about printing? Would it make us more cautious? How would we balance this with the wish to publish a broad spectrum of material?

There are different types of printing: black-and-white web-fed offset litho, color sheet-fed offset litho, digital short-run inkjet, and so on, and there are different ways of selling into the marketplace—perhaps your business uses a co-edition printing model or an academic publishing model—but in all these scenarios it is commonplace for the printing of a title to take place on the largest print run that seems appropriate based on projected sales. A large print run has often been aimed for, as traditional litho printing requires a make-ready to set the print machine up and ensure ink coverage is correct, and the unit cost will be lowered up to a certain point as the make-ready per unit is reduced with the costs spread across a larger quantity of books. The longer print run has supported the publisher and retailer goals of keeping titles continually in the warehouse. But is the longer print run, which can incur transport costs, storage costs, and over-printing costs, genuinely worth it financially? Many are now questioning this and are finding ways to reduce working capital.

SOLUTIONS: JUST IN TIME

Different print methods, print-run lengths, and responding to new sales patterns and requirements will all need to be addressed to find ways of reducing returns in the publishing industry, whether it's a small print run of a black-and-white book as part of a publisher's back catalog or a highly popular color title.

Key to reducing returns will be to ensure that titles can get quickly to market, making it clear within your business internally as well as to your retail customers that this will be possible. And this will be made possible by developing new and additional print strategies and distribution models.

• **Print locally.** This speeds up the delivery time, with potentially a very short time between the order being placed with the publisher to the order being delivered to the customer. Printing locally reduces the need on the publisher's part to store books and the need of the bookseller to order in bulk, thereby potentially over-ordering, because all parties will be aware that they can have books dispatched quickly if required. The lead time could be reduced from twelve weeks to one week or less depending on the printing method. The cost of printing locally may be more at present but, with rising oil prices increasing the cost of transport, these costs will start to even out, making printing locally increasingly worthwhile. Printing locally reduces returns and associated costs, as well as transport and storage costs.

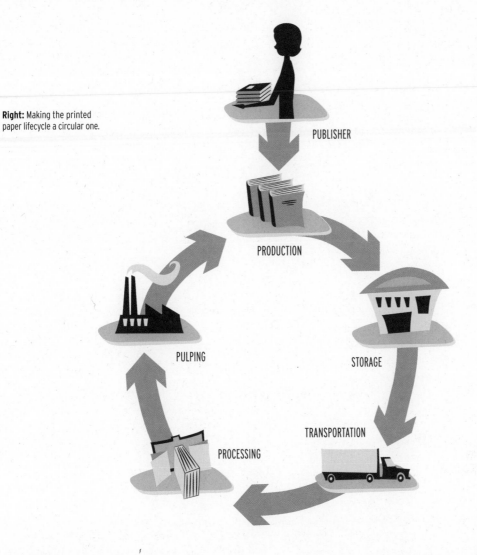

Right: Making the printed paper lifecycle a circular one.

PUBLISHER

PRODUCTION

STORAGE

TRANSPORTATION

PULPING

PROCESSING

- **Digitally print short runs.** Lower storage costs and reduce the need to have cash tied up in stock by regular short-run printing. Digital printing reduces paper and ink use, as there is no set up required, and represents an environmental bonus in this respect. Printing digitally is also making short-run printing for backlist titles a more viable financial option—previously, the make-ready proportion against a short print run had made the unit cost of a book printed on a short-run length too high. If combined with printing locally, there will be the dual benefits of lower transport costs and less need for storage either with the publisher or with the bookseller.

- **Digital print on demand (POD).** Print on demand can open up a publisher's back catalog and reduce the need to have out-of-print books. A substantial part of a publisher's costs are in content creation: opening up the back catalog and printing previously out-of-print books, or repurposing existing content, re-monetizes that content. And it also probably makes your customer base very happy too as they can again find their niche title. The print-on-demand principle and the shorter print run may well also be applicable to magazine and journal publishing as a way to reduce overprinting. This is a practice already adopted by smaller journal publishers, and it may work well with a subscription

CASE STUDY: WILEY

Wiley is a wide-ranging academic publisher (CliffsNotes, postgraduate materials, etc.), and has this to say about its use of digital printing: Wiley is leveraging digital initiatives to respond to customer demands regarding content delivery and format. We work with local digital print providers to optimally manage our inventory, micromanage supply, order smaller volumes more frequently, and continually adjust to market demand. This aligns with both our business strategies and our Corporate Citizenship goals, and presents an opportunity to decrease our carbon footprint. For example, in the UK, small printings are consolidated at our European Distribution Center to provide consignment shipping to customers. In some cases in the U.S., we ship directly to the customer from the print vendor. We realize significant supply chain efficiencies and carbon savings from both models, particularly as the program greatly reduces the need for transatlantic shipping.

model. It could be thought of as an extension to the subscription model, as a first print run could be for subscribers and later print runs would be in small batches or on demand, depending on volume of demand. The risks associated with overprinting are reduced.

- **Short-run printing.** Depending on your product, the color quality of digital printing may not be suitable at present, but developing short-run litho printing programs with existing suppliers sets your business model moving towards keeping lower stocks. There is no immediate environmental benefit, such as reduced material usage, but shifting a business model and printing patterns is a large task and all moves towards lower stocks are positive.

- **Distribution.** Changes to print strategies will require corresponding changes in distribution, which, to achieve the needs of the market, as well as to be as environmentally friendly as possible, will need to be small, efficient, and utilize consolidation. The consolidation of deliveries to avoid partially full containers is an environmentally sound practice, that also makes good business sense, and is an important part of distribution management as it is already practiced. The change will come about through moving from an emphasis on global distribution networks via a central hub towards an emphasis on highly evolved

local distribution networks. For example, instead of consolidating the output of, perhaps several, printers from an Asian port, delivering to a central hub, and sending out to individual customers, the consolidation will take place at the local printer or the local distribution hub. This will result in fewer book miles, allowing the main emphasis of management to be on reaching the customer as quickly as possible rather than on managing a large global network. Consolidating from your local printer means that distribution becomes smaller and more efficient.

- **Digital readers.** In conjunction with POD and short-run printing, releasing titles to be read digitally may well prove useful for backlist titles, as a way of monetizing previously out-of-print content, or for titles that are released with regular updates. And while reading digitally has its own environmental impacts, this can nevertheless be a useful part of the solution for getting titles to customers in a way that reduces overstocks and returns.

These solutions work well in conjunction with each other, reducing costs in several areas at once. These initiatives can also be released gradually. You will have different products selling into various parts of the marketplace, all with their own purchasing patterns, and where there are different purchasing patterns it is likely that different combinations of the suggested solutions will be needed.

Right: Digital readers can play a role in getting content to customers as efficiently as possible.

For instance, if a backlist title is incurring storage costs and returns, you may want to print it digitally on a short run. And if there is a trend for frontlist titles to incur returns, then you can make an initial print run of optimum length to reduce your unit cost (based purely on the minimized make-ready) and then print reprints locally and frequently for quick delivery into the marketplace. As deciding factors of cost shift with changes in oil prices, it is likely that print and distribution strategy changes will begin gradually then pick up momentum.

In addition to changes in print and distribution strategies, changes in the stock management relationship between publisher and bookseller could help reduce returns. Some of the following options may be suitable for your business relationships with retailers:

- Review reordering cycles
- Review minimum ordering quantities
- Reassess promotional strategies and concentrate on sell-through rather than purchase
- Can a markdown be agreed on a title rather than incur the returns cost?
- Consider firm sale on deep backlist

The options considered may initially incur increased unit costs, and it is likely that these costs are already well known, and it is to avoid them that other print strategies have been used. However, if used effectively, they can also reduce costs in many areas, such as distribution miles, storage, over-production, returns management, and disposal.

If these cost reductions can be captured and included in print decisions, this will be an excellent guide to the ways in which the different print and distribution strategies can be utilized, thereby reducing the financial and environmental burden of returns, while ensuring their original function—which is to promote diversity of publishing.

Reduce, reuse, recycle

Returns will never be entirely eliminated; after all, they serve a useful function in that they ensure the ongoing diversity of publishing. This means that solutions need to be found for appropriate product disposal. That well-known mantra "reduce, reuse, recycle" can help us think about what to do with books at the end of their life. We have looked at the first aspect, "reduce," as we aim to reduce the number of products made and then returned to the publisher. We will now look at ways that products can be reused or recycled.

To avoid books becoming an additional environmental impact, putting them into landfill should be avoided. All paper-based products contain the carbon stored by the trees they were produced from, and this carbon storage will be negated if they enter landfill and decompose, thus releasing the stored carbon. If books enter landfill, they are also contributing to the increasing problem of landfill site availability. Better to reuse and recycle them. We will look at reuse and recycling in the context of the industry (publisher and retailer) and then at the point of consumption (publisher and customer).

Right: Reuse requires no energy, and benefits additional readers.

PUBLISHER AND RETAILER

Reusing products has the potential to be the most energy and financially efficient option available to a business. The management of stock can play an important role in reducing the stock returned and then in managing the returned stock for use elsewhere. This is a well-known problem, and the UK's Industry Returns Initiative has identified some solutions in this area (and returns have reduced in the UK). Items for return could be streamed into "Red box" and "Green box" return receipts. Green receipts would be those items that could be easily resold if returned within a particular timeframe and in good condition. Titles returned within a particular timeframe after publishing may well be easily sold elsewhere in the business without excessive time or expense being required to do so. Where categories can be agreed and adopted by the industry, this becomes a simple system to operate, with benefits on all sides. The returns system should be streamlined for ease of use and clarity, with an emphasis on either return for reuse or disposal.

Once it has been decided that a book cannot be sold and should no longer be stored, it must be disposed of. And one way in which it can still be reused in disposal is as a charitable donation. There are many organizations that would benefit from books or journals, but who wouldn't have the funds available to purchase them.

The second disposal option is to pulp and recycle them. For retailers and publishers, the case can be made that both could pulp and recycle books and paper products.

If agreements can be put in place that allow books to be disposed of onsite by the retailer, instead of the traditional return, it would reduce an aspect of the returns management system and save greatly on book miles: instead of returning the book, a "Red box" receipt would be sent, which is a feature of the system as proposed by the Industry Returns Initiative mentioned previously. The decision about which titles could be disposed of this way would be part of the agreement between retailer and publisher. Simplifying the system and making it more environmentally friendly are intricately linked.

GREEN RETURN RECEIPT	RED RETURN RECEIPT
Title to be reused and returned	**Title to be disposed of**
· Title is in a returnable category, e.g., not a firm sale	· Title is x months post publication and y months since order date
· Title is x months post publication and y months since order date	· Title is damaged and unfit for use
· Title is in good condition	· Title to be disposed of by retailer or publisher
· Agreements in place between publisher and retailer for the specifics of category, age, condition, site of disposal, and so on	

Above: Books have great recycling potential as there are clear points at which the material can be collected.

One of the issues to consider in relation to recycling is this: what happens to the reclaimed material? Publishers face an increasing pressure on materials, with deforestation causing major environmental impacts. While publishers are not manufacturers of paper, they are a part of the paper manufacturing supply chain. Is there room for the publishing industry to help create a closed loop recycling system in which returns can be pulped and sold onto the paper recyclers, guaranteeing both the quality level of product and an increase in the level of recovered material available in the paper market? The examples of where this has been most effective come from manufacturers

such as Interface. Interface is one of the largest global manufacturers of carpet and is on a mission to reduce its environmental impact to zero. Carpet manufacturing can be a highly oil-intensive process, and an important part of reducing the environmental impact has been the use of recycled materials; since 2007 this has included a recycling of yarn process from post-consumer material. The use of recycled materials has reduced the company's cost base as the overall manufacturing process, including recovery and processing, is now less energy intensive.

Some commentators have suggested the idea of "leasing" as a way for industries to reduce the environmental impacts of their products. A product is "leased" for use to a customer at a fixed cost, perhaps a one-time-only cost, or it could be a monthly fee, and at the end of the product's lifespan the materials are reclaimed by the manufacturer or its supplier and a new product resupplied to the customer. In publishing this would not be an entirely new process as there is already pulping for recycling, but it is a process that could be built up and expanded as a way to reduce costs and resource use. It would require a radical revision to the costs and supply model, but as publishing is changing fast, and these models are changing too, now is a good time to explore options that respond to long-term environmental and financial concerns. Or alternatively, could partnerships be formed with organizations who are given books when customers have finished with them and who may have excess or unsaleable books, such as charities and thrift stores?

CUSTOMER AND PUBLISHER

Customers can recycle books and paper products through locally provided recycling facilities (often a combination of curbside collections with centralized recycling points). Customers also commonly send books to charities, or they enter the secondhand market. However, not all published products are solely paper-based. Advice to the consumer on how to dispose of non-paper based products would be helpful, and the publisher may also be required to create provision for recycling and disposal in some instances.

Many publications now come with component parts that may, or should be, recycled. In the children's book market sound modules are commonplace: batteries within sound modules can leak dangerous chemicals into landfill and in the UK, for example, if you supply more than 70.5 pounds a year of batteries to customers or business, you are required to take back those batteries from customers, free of charge, for appropriate disposal and recycling. You may not consistently supply over 70.5 pounds of batteries, and it may be appropriate to direct your customers to sites where batteries are commonly collected for recycling, such as supermarkets. But if you are exceeding the threshold each year, setting up a recycling scheme in conjunction with retailers could prove a suitable solution. Even if you are operating in a country without a comparable regulation, the response to the UK regulation can provide a useful model for businesses to manage the environmental impact of batteries and avoid batteries going to landfill.

Some publications will also have a complex mix of component parts that make them hard to recycle. To reduce landfill it would help customers to have information on how to dispose of these products: instead of throwing these items away, they could either be separated into parts, for instance cloth, plastic, and magnets, and sent to local recycling sites, or more simply (and with a lower environmental footprint), they can be donated to charities who may sell them on or give them to local toy libraries, hospitals, and so on. This advice could usefully be put on a publisher's website.

"Customers are now increasingly well informed about the environmental impacts of the products they purchase."

We see that there are three main points at which publications could be disposed of: by the retailer instead of being returned, by the publisher if the return was unavoidable or appropriate at the time, or by the customer after purchase. Each of these groups has to be enabled to dispose of products appropriately, and the publisher and retailers can play a key role in this. Agreements between retailer and publisher can be put in place to streamline the returns process and allow retailers to dispose of items on site rather than going through a lengthy and higher-energy returns process. Publishers can recycle much of the returned products and look at ways to maximize the value of this raw material; they can also provide information to customers on how products should be disposed of, particularly whether or not they are suitable for recycling, and other disposal options if they are not. In this way a new form of lifecycle could be encouraged, moving away from the linear product cycle in which materials are taken, used, and disposed of, and moving towards production, use, disposal, reclaiming materials, production: a circular lifecycle.

CUSTOMER ENGAGEMENT

Customer engagement is one of the key issues for the publisher after the production of the books. In customer engagement we encounter both issues of the customer's role in the environmental impact of a product through disposal methods, and also the issue of what the customer expects from the publisher—now and in the future. So the question arises: how should customers be engaged with on environmental issues now? Are they a sleeping giant who will wake up and dictate how publishers operate at the environmental level? In a sense, yes, the relationship between publisher and customer is changing. It is not a simple interaction directly between publisher and customer; it includes input from the press and environmental NGOs, and also the perception of a product category that sets the customer's expectations before establishing a specific relationship between an individual customer and a publisher.

Customers are now increasingly well informed about the environmental impacts of the products they purchase through traditional media, represented by investigative journalism, and through the viral campaigns of environmental NGOs. Environmental NGOs are initiating a circular relationship between NGO, journalism, and customers by narrating the environmental issues that they are making public, allowing the public to more fully engage with environmental issues—from palm oil production and its role in deforestation to the atmosphere-polluting role of data centers if they don't use low carbon electricity. They are also creating a hook that can be used by a journalistic article, and there is interesting interplay between traditional media and environmental NGOs and their use of social media, with which traditional media compete for attention. This process of narrating has allowed consumers to see how their choices play a part in environmental impacts, which

Right: Find locations where customers can dispose of products, even when they are complex, and communicate this to them.

has engaged them further, but the emphasis of the campaigns is often, correctly, on the companies that are responsible for the activities creating the environmental impacts. While consumers are increasingly well informed, they are not in all instances putting that knowledge into practice when making purchasing choices.

Whether or not customers choose to bring their knowledge of environmental issues into play when making purchasing choices seems to take in several factors: the product category, the content, and the company that produced the product. Books as a product category are trusted, people like books, they contain knowledge, and are thought of as a positive aspect of culture, and publishers are trusted as the makers of good books to make them in a good way, to be doing the right thing in terms of the trees used in making paper and in the ethics of their production. Books are also viewed as simple products because they often only have two obvious and visible component parts, the text and the cover, though here a product that looks simple often belies a product with a complicated supply chain.

Children's titles, particularly the more complex ones, have a different set of assumptions attached to them. A title's complexity, and its close resemblance to toys, can psychologically move it into the toys category, and here a consumer will start to have more concerns about the way in which a product was made, by whom, and under what circumstances. These associations and the desire to purchase items for children that have not had a negative environmental or social impact create a higher set of expectations for the consumer, and also a greater desire to make sure that these expectations are met, unlike the more passive expectations associated with other publishing categories.

How can these different expectations be met by the publishing industry and by individual publishers? How can publishers respond to customer expectations and increasing levels of knowledge, and where should the publisher fit into the network of information and debate around production?

Green claims

Above: An example of greenwash; unsubstantiated, or as in this case, uninformative claims.

Green claims on and off product, while not yet common, are seen with increasing frequency, and there are positive reasons for using them as a response to your customers' concerns and interests. By reducing the environmental impacts of your products and letting your customers know what you are doing, and why, you can expand your customers' knowledge about what the publishing industry's environmental issues are and add to the debate about what can be done and how it should be done. Although this role is often occupied publicly by journalism and environmental campaigning organizations, businesses that take an active role in reducing environmental impacts can make a very positive contribution to debates by bringing their specialist knowledge and experience of practical resolutions to these issues. This is also an opportunity to let customers know that they have a role, beyond their initial choice of your product, in reducing the environmental impact of the product by minimizing the impact of disposal at the end of the product's life.

The UK's Department for Environment, Food, and Rural Affairs (Defra) has produced some guidance that is useful for the initial framing of all our claims.

Defra's summary green claims code states:

1. Ensure the content of the claim is relevant and represents a genuine benefit to the environment
2. Present the claim clearly and accurately
3. Ensure the claim can be substantiated

This green claims guide helps remind us broadly about what the customer will want to know, and what we will need to know to back up the claim. It helps us avoid greenwash (the appearance of a greater level of environmental responsibility than is actually present, such as suggesting paper is green solely because it is natural). Greenwash can derail industry efforts at transparency and create negative connotations in the product categories we are working with.

ON-PRODUCT CLAIMS

Logos or statements need to be specific and accurate to that particular product. For this reason, a lot of work has often gone into producing a logo or statement that can be put on a product, but not all logos or statements are equally recognized or perceived as being meaningful. Studies have shown that the more familiar a logo or phrase, the more meaningful it becomes to the customer. This idea also fits with the fact that customers make purchasing decisions in a relatively short space of time, weighing up several factors. As environmental issues may not be one of the customer's factors, the logo or statement should be easily grasped, and digested; that is, it should be quickly meaningful. The information conveyed on a product must therefore be comparatively limited, and providing a link to a further source of information for anyone interested is useful.

OFF-PRODUCT CLAIMS

Online you can talk about all your organization's activities, policies, achievements, and goals. Providing more in-depth information about the environmental issues of publishing, your company's efforts to reduce its environmental impacts, and the publishing industry's position creates transparency about your business and the industry it works within. It also furthers the debate about what best practice should be in these often substantial issues.

Furthermore, it's an opportunity to involve your customers beyond their product choice and show them how their actions—the use and disposal of a product—can have an environmental impact that can be reduced and kept to a minimum. In this way the customer can be included positively in the notion of responsibility: the responsibility for the environmentally sound production lies with the publisher, who demonstrates that they are living up to that responsibility, and responsibility for the use and end life of the product lies jointly with both publisher and customer, with the publisher supporting the customer in choosing a disposal option that keeps the environmental impact to a minimum.

These two approaches combined inform your customer about your business' efforts to become greener. When you have reduced your environmental impacts, letting your customer know about it shows that the faith they placed in your products was well placed, and it brings your business into the debate about which business practices are best for the environment at a time when all our efforts are needed to reduce the environmental impacts of industry.

Above: Combine verified on-product claims with in-depth information online.

We have seen in this chapter that a publisher's responsibilities to lower environmental impact extend beyond production of a product and into the use and end-of-life phases of that product. All efforts to reduce, reuse, and recycle by working with your suppliers and retail customers on print and stock management, and through customer communication, will lower the environmental impacts of your products and the publishing industry as a whole.

In the next chapter we will look at future challenges for the publishing industry and where else we might lower environmental impacts.

10

The future is to be written

Publishing, and the supply chain it depends on, is changing: we see increased regulation and changes to product labeling globally, along with its consequences and impacts; we see changes in the technologies used to deliver the published content; and we see developments in the response to resource material pressures. And as all these changes occur we need to address their impacts and incorporate them into our business thinking. Looking to the future, the challenges publishers face and how they might respond to them show not only the breadth of issues, but that there are already mechanisms in place that will help address them.

Increased regulation

This evolving process can be seen around us. In Chapter 1 we looked at the drivers for change in the global context, which included important pieces of legislation such as the U.S. Lacey Act and the EU Timber Regulation (EUTR). The EUTR does not currently include imports of printed paper, meaning that much printed material sold in the EU will currently be outside the EUTR. But printed material is a potential inclusion for the future, particularly if it is shown by the forest certification NGOs and the EUTR monitoring bodies that the systems exist to document the origin of the pulp fibers contained within a paper, and that therefore it is possible to regulate paper beyond its manufacture, so that those involved in its purchase can take responsibility for that purchase.

The regulation and voluntary frameworks around product labeling, in particular for carbon, also look likely to be extended. This will have implications for publishers and their supply chains globally. Carbon labeling of products is already with us to an extent, with supermarket chains in the UK and in France even now labeling some own-brand products. At present the number of customers who look for a carbon label is small, and few as yet understand it, but product labeling is gathering pace as a movement, however, as it provides another important benefit beyond communication. The process that a business goes through to assess a product's carbon or environmental footprint highlights the opportunities for reducing those footprints by quantifying and making transparent all the environmental impacts of a product.

It has become clear that not all product labels are comparable: for example, the carbon labels of the French retailer Casino do not include a "use phase" and, depending on the nature of the product, this can be an important carbon emission-producing part of the product lifecycle. Electronic products will clearly have an important use phase as they use energy directly; clothing has a less obvious use phase, but necessary washing does require energy and other material inputs (cleaning materials); bananas do not typically have an energy-using use phase. For this reason the most accepted carbon-accounting methods and labels do have a "use phase" aspect (PAS 2050, ISO 14067, GHG Protocol) and this use phase will be utilized if the Product Footprint Category Rules (PFCR) of that product have defined that it should. PFCR are agreed across an industry to ensure that footprints are meaningful and comparable.

France has begun a lengthy experiment in product labeling with 168 companies signed up. The products produced by these companies will have footprint labels, certainly for carbon, and also for other factors such as biodiversity and water, depending on the PFCR for that product. The PFCR have been defined by both the French Ministry for Environment and AFNOR, the French Standards Agency. This is a year-long experiment and is thought to be a precursor to mandatory labeling for products on sale in France and would consequently affect all imported products. Hachette began labeling their books sold in France in May 2012. This national strategy

Right: A breakdown of factors in the publishing industry's environmental footprint.

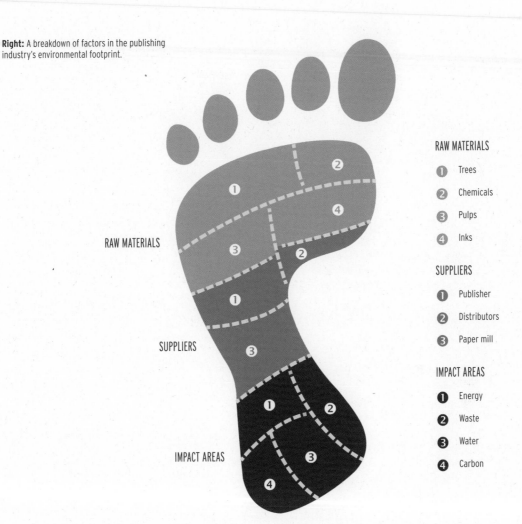

RAW MATERIALS

1 Trees

2 Chemicals

3 Pulps

4 Inks

SUPPLIERS

1 Publisher

2 Distributors

3 Paper mill

IMPACT AREAS

1 Energy

2 Waste

3 Water

4 Carbon

will have cross-border impacts; the effect already has been to increase the interest in a Europe-wide footprint labeling system, and it has highlighted the need to make labels and footprinting methodologies comparable.

It is the importance of comparability that has prompted the European Commission and CEPI to begin defining PFCR for paper as part of a pilot on the development of PFCR. A widely accepted PFCR for paper that assesses the full supply chain—from cradle (the raw materials, trees; through materials processing, chemicals, pulps, and so on) to grave (paper mill gate)—could point to ways to save energy in paper manufacture. If paper products are sold with footprint labels this could, in turn, present an opportunity to significantly reduce the environmental impact of printed products. An established and recognized footprint methodology for paper will also make it easier to create a footprint for printed media, which would incorporate a use and disposal phase.

Digital media

The knowledge of the publishing industry about our environmental impacts and how to account and take responsibility for these will evolve and expand. This is highlighted already by the development of accounting and labeling methodologies and will be an ongoing feature as we incorporate digital products into our businesses. If we are producing content for a digital reader, where do our responsibilities lie in terms of environmental impacts? If we account for our business impacts when producing content (GHG scope 1), do we then account for the "use phase" while a product is used, for example when a book or magazine is read by digital means? We can account for this phase, as can the manufacturer of the reader, but who then takes responsibility for the associated impacts? Can publishers find ways to encourage readers to read in the least impactful way possible, for example downloading content rather than reading online, while still delivering the high quality product the customer wants? There are existing frameworks that can guide thinking about these issues, but there is still a need for more knowledge and for agreement on accounting and reporting scopes. Boundaries of responsibility, or even shared responsibility, will be clearer once the supply chain is fully known and understood—from manufacture of devices through to creation of content and end use—and then all parties can work together to reduce impacts.

Right: Digital publishing will reduce material impacts and at the same time add new environmental impacts to be managed.

Resource pressures

As populations and consumption of goods increase, the demand for timber grows. The world's natural forests cannot support this alone. We are able to monitor the forest sources of our papers now with increasingly sophisticated risk-assessment systems, and therefore we are able to source paper responsibly, but demand for fiber is increasing, and those wood fibers have to come from somewhere. Knowledge of the sources cannot be the sole answer, then; there must be additional responsible sources of fiber. There are certified sources, as we have seen, and also recycled material. Are there ways for these two categories to be increased? To increase the recycled material available would require a change in the marketplace chiefly; there would need to be a higher recovery level and a greater demand for the papers.

An increase in certified sources would perhaps be well answered by reforestation. Deforestation and land-use change have left many areas degraded, without benefit to the plants, animals, and people that have relied on them. Reforestation that goes beyond mono-crop plantations could provide long-term sources of fiber and livelihoods that support both local needs and bigger businesses. There are many types of forests and there can be many goals of reforestation, and so reforestation can take many forms: areas set aside for natural regeneration, the planting of native tree species, and agroecology.

Such projects could again provide a living space for people, plants, and animals, and some of these projects could also reduce the need to log natural forests. With up to 20% of humanity directly dependent on forests, this could be a truly worthwhile activity. Certification systems are reviewing this area, and it would be immensely positive if reforestation could become a regular practice, and if it could be incorporated into certification in the same way that set-aside (the practice of leaving a portion of forest area un-logged) is built into certification systems and thereby supported by all those who use certified materials and products. What this pressure on resources indicates is that the problem all business faces is the problem of growth. Our economic, and indeed social, model is predicated on continual growth, which we have had in the past, and today we see the rise of both consumption and population, which creates this growth. But it is precisely these rises in consumption and population that are putting pressure on the world's resources and destabilizing our environments, our climate, and our livelihoods.

But growth has been necessary; without growth, output falls, as does a company's financial return, putting downward pressure on jobs and investment. How then do we resolve this substantial problem?

Maintaining employment and product quality in the face of, or in preparation for, reduced output may require

"Indefinite material growth on a planet with finite, and often fragile, natural resources will however, eventually be unsustainable."

(BLUE PLANET PRIZE LAUREATES—ENVIRONMENT AND DEVELOPMENT CHALLENGES: THE IMPERATIVE TO ACT, 20.02.2012)

evolution of hours and pay structures across the board. Such ideas are embryonic and still in the process of being worked through, but this evolution can certainly be prepared for by reducing material requirement proportional to product output, and many of the strategies proposed in this book have that goal in mind. Reducing material requirements (paper, energy, and so on) reduces the cost base and the environmental burden. There is no complete answer to this yet; we are in a time of transition. But the beginnings of the answer lie in the strategies gathered in this book.

Preparing for the future

We are looking for a reduction in material resource use in absolute terms, while at the same time maintaining employment and other business functions; in the publishing industry's case that would be the published content. We will see developments in regulation that will affect publishers and their supply chains, and that will need new frameworks, or extensions to those that already exist. We will see an evolution in our understanding of how our activities affect the environment and the best practice to minimize that impact.

The process and methods suggested in this book will help prepare for all these changes and challenges.

In the publishing industry we see the beginnings of a cultural shift, with ideas being shared and often worked on collaboratively to help each other achieve the goal of lower environmental impacts and to more effectively reduce the impact of the overall industry. And these ideas show that the way business understands its role in the world is starting to change and become more integrated, in a positive sense, into the environment and society rather than imposing itself on top of it.

The ideas here are based on well-known precepts and have been further developed by many people into highly workable strategies for companies to employ and reduce their environmental impacts. These tools and methods have been gathered here to give you what you need to make a difference in your organization, to show that keeping the environment at the core of all your business decisions makes sound economic sense, to reduce your environmental footprint, and to set you on the path to zero impact.

PROCESSES AND METHODS FOR CHANGE

- Adopt a green supply chain model
- Carry out a Lifecycle Analysis (LCA) of your products
- Reduce impacts based on that knowledge
- Extend product responsibility from cradle to grave
- Gather information for labeling and communication both business-to-business and to customers
- Change your business culture
- Through green teams and green projects, show how best economic practice for business is fundamentally linked to what is right for the environment and society
- Value the environment for its ongoing provision of homes and livelihoods, for supporting an incredible diversity of plants and animals, and as a cultural cornerstone
- Implement a business-wide EMS
- Use the EMS to plan, monitor and record your knowledge-gathering and reduction efforts
- Use the information gathered to communicate business-to-business and to customers
- Effective and responsible material use
- Design products for minimized material use
- Know the sources of the materials you use

- Use certified materials
- Detoxify your supply chain through your material choices
- Make print decisions that minimize material use
- Increase material recovery levels in your own business and through communication with customers
- Close the loop and use recycled materials where possible and increasingly
- Monitor and aim for a reduction in material use in relative terms, for example, lower resource use relative to output, and aim for absolute reductions with overall less resource use
- Green printing, localization, and resilience
- Choose printers with environmental accreditations, strategies, and proven environmental impact reduction
- Print locally and often; reduce transport, storage, and overprinting costs
- If printing overseas, aim to deliver direct and not via an interim warehouse to reduce book miles
- Make your business more resilient by making it less resource dependent (less virgin raw material and energy dependency in particular)

References

Appendix

1. An explanation of the different ways in which publishers and printers refer to paper weights, and how to convert from one to the other

There are two systems in use to express the weight (also known as *substance*) of a sheet of paper.

One is to express the weight of the paper in grams per square meter, written either as gsm or g/m²: for example, 90gsm or 90g/m². The paper's weight is sometimes called its *grammage*. Book paper weights start at about 27-30gsm for lightweight or Bible papers and go up to about 110gsm before they become either too heavy or bulky to be handled easily during mass production. Generally speaking, paper for book work falls within a fairly narrow range starting at about 70gsm and going up to 100gsm in increments of 10 grams. The final choice will be dictated by the nature of the book and the processes used to produce it.

When choosing a paper, it is worth remembering that:
• heavy papers make for heavy books
• opacity does not necessarily increase with weight; the book just gets heavier
• thick or bulky papers are not necessarily heavy
• bulky papers make for bulky books
• bulky, heavy books are more expensive to transport and store than thin, light books and therefore have a correspondingly larger environmental impact

The other system, which is predominantly used in the USA, describes paper weight by *basis weight*, which is the weight in pounds (lbs.) of a ream of paper (500 sheets) cut to its basic size. The basic size is determined by the grade of the paper. For example, the basic sizes for the following grades are:

PAPER GRADE	BASIC SIZE (IN INCHES)
Cover boards	20 x 26
Newsprint	24 x 36
Book papers	25 x 38

A book paper described as "60 lb." (sometimes written 60# basis) means that 500 sheets of that particular grade weigh 60 pounds, and this would be written as: 25 x 38 - 60 (500).

To convert the basis weight of a book paper (25 x 38) to gsm, multiply the basis weight by 1.48. So, the 60 lb. book paper in gsm would be 60 x 1.48 = 88.8gsm. To convert gsm to lb. basis for a book paper, multiply the gsm by 0.675. So, a 70gsm paper would be 75 x 0.675 = 50.6 lb. paper.

Paper can be sold in units of 1,000 sheets, in which case the basis weight for a book paper would be expressed as: 25 x 38 - 120(M). In this example, all you need do to work out the basis weight per ream of 500 sheets is to divide the 120 by 2, which will give you 60.

For *rough* conversions of *book papers* either way,
remember these two factors:
• **lbs. to gsm:** multiply by 1½ • **gsm to lbs.:** multiply by ⅔
Full conversion tables for all grades of paper appear below:

BASIC SIZE (INCHES)	FROM GSM → LBS. MULTIPLY BY	FROM LBS. → GSM MULTIPLY BY
17 x 22	0.266	3.76
20 x 26 (boards)	0.370	2.70
20 x 30	0.427	2.34
22 x 38	0.438	2.28
22 ½ x 28 ½	0.456	2.19
25 ½ x 30 ½	0.553	1.81
23 x 35	0.573	1.75
24 x 36 (newsprint)	0.614	1.63
25 x 38 (book papers)	0.675	1.48

2. Book sizes
In the UK, common book trimmed page sizes are
measured in millimeters, with the height first then the
width. This produces portrait-format books. Sizes are
given a name, which appears in the left-hand column.

NAME	TRIMMED 8VO IN MM	TRIMMED 4TO IN MM
A format	178 x 111	-
B format	198 x 129	-
Crown	186 x 123	246 x 189
Large Crown	198 x 129	258 x 201
Demy	216 x 128	276 x 219
Royal	234 x 156	312 x 237

In the USA the common trimmed page sizes are measured
in inches, with the width first followed by the height,
to produce a portrait-format book.

SIZE IN INCHES	EQUIVALENT SIZE IN MM
5 ⅜ x 8	137 x 203
5 ½ x 8 ¼	140 x 210
6 x 9	152 x 228
6 ⅛ x 9 ¼	156 x 235
7 ½ x 9 ¼	190 x 235
8 x 10	203 x 254
8 ⅜ x 10 ⅞	216 x 279

3. Some useful formulae

Calculating the Weight of a Book

This is useful for working out things like:

- postage rates
- the number of books you can include in a binder's parcel (where the weight is limited to a maximum of 29.7 lbs. (13.5 kg))
- the number of books you can load onto a pallet (where weight is limited to 1 ton)
- shipping tonnage

The formula is:

page size (mm) x paper weight (gsm) x half the extent/ 1,000,000

For a paperback, add 20g for the cover; for a hardback, add 100g for the case.

Example:

What is the weight of a 320-page paperback book in a format of 216 x 138mm, printed on a 90gsm paper?

$$\frac{216 \times 138 \times 90 \times 160}{1,000,000}$$

Book weight is: *429g + 20g for the cover = 449g*

Remember to use *half the extent* as this is the number of *sheets* in the book, each sheet carrying two pages!

If you are in the USA and working in pounds and inches, calculating the weight of a book is done differently. As we already know, the basis weight of a book paper is expressed as so many pounds per ream of 500 sheets in a standard size of 25 x 38 inches: for example: 25 x 38 – 60 (500). Where your book is printed on a paper of this weight and size, calculating its weight is fairly straightforward. You need first to calculate the number of sheets required for one book, and then multiply the result by the weight of each sheet. This is shown in the worked example below:

A 320-page 6 ⅛ x 9 ¼ book, printed on a 60 lb. 25 x 38 sheet requires 10 sheets of paper.

To calculate the weight of each sheet in the ream, divide 60 (lbs.) by 500 (sheets) = 0.12 lb. per sheet. So, 10 sheets of this paper (i.e., the book block) will weigh 1.2 lbs., to which should be added ¾ oz. for a cover, and 3 ½ oz. for a case.

However, not all books are this size, and we need to know how to calculate the weight of a book that is not printed on a 60 lb. 25 x 38 sheet.

For example, a 5 ½ x 8 ½ book is most economically printed on a 35 x 45 sheet. How much would a 192-page 5 ½ x 8 ½ book weigh if printed on a 50 lb. 35 x 45 sheet?

To calculate the weight of the ream, use the formula:

$$\frac{basis\ weight \cdot x\ sheet\ dimensions}{basic\ sheet\ dimensions}$$

For the paper above this would be:

$$\frac{50 \times 35 \times 45}{25 \times 38} = \frac{78750}{950} = 82.89\ lbs.$$

To calculate the weight of each sheet, divide the ream weight by 500:

$$\frac{82.89}{500} = 0.16578$$

Each book requires 3 sheets x 0.16578 = 0.49 lb. or 7.8 oz.

So, the weight of a book is just under ½ lb.

Calculating Paper Quantities

Sheet-fed work

The formula for calculating how many sheets of paper are needed for a job is:

$$\frac{extent\ x\ print\ run}{pages\ printed\ on\ a\ sheet}$$

Example:

How many sheets of paper will be needed to print 1,200 copies of a 320 page book, if printed 16 pages per sheet (or 8 pages to view)?

$$\frac{320 \times 1,200}{16}$$

The number of sheets needed is: 24,000

To this quantity you need to add spoilage; the spoilage rates appear below.

Web-fed Work

Again, you need to know the extent and the print run, as well as the number of pages carried on each side of the paper.

But paper for web presses comes on reels, and not in sheets, so you have to work out the sheet size based on the reel width and the size of the press cut off, which is what cuts the printed reel into lengths.

With this information, you can work out the number of pages printed to view in exactly the same way as you would for a sheet.

Example:

A 320-page 234 x 156mm book, printed on a 960mm wide reel on a press with a 640mm cut off, ends up as a 10 x 32-page folded sections (printed 16 pages to view).

If each section uses 640mm (or .64 of a meter) of paper, the total amount needed for the book is:

10 x 0.64 = 6.4 meters

35,000 copies of the book use 6.4m x 35,000 copies = 224,000 meters, plus spoilage

If you are working in inches, feet, and yards, the principle is exactly the same:

A 320-page 6 ⅛ x 9 ¼ inch book, printed on a 38" wide reel on a press with a 25" cut-off, ends up as 10 x 32-page folded sections.

Each section uses 25" of paper, so the total needed for each book is 250" or 6.944 yards.

35,000 copies of the book use 6.944 yards x 35,000 = 243,040 yards

Spoilage Allowance Rates

Spoilage is the term used for the paper lost or wasted during printing and binding. The figures below show typical figures for spoilage, expressed as a percentage of the paper needed for the print run. Spoilage allowances vary according to the number of copies being produced and the nature of the work; the exact amount of spoilage is something that needs to be discussed and agreed with the supplier early on in the life of the project, since in the end it is the customer who pays for the extra paper.

For monochrome sheet-fed work, typical spoilage allowances are: (*For 4-color work, add 2% to the printer's allowance.*)

PRINT RUN	PRINTER'S ALLOWANCE	BINDER'S ALLOWANCE	TOTAL ALLOWANCE
1,000-2,500	4.0%	2.5%	6.5%
2,500-5,000	3.0%	2.0%	5.0%
5,000-10,000	2.5%	1.5%	4.0%
10,000 and above	2.5%	1.0%	3.5%

For monochrome web-fed litho work, typical spoilage allowances are: (*For 4-color work, add 3% to the printer's* allowance.)

PRINT RUN	PRINTER'S ALLOWANCE	BINDER'S ALLOWANCE	TOTAL ALLOWANCE
10,000-15,000	11.0%	1.0%	12.05%
15,000-25,000	9.0%	1.0%	10.0%
25,000-75,000	7.0%	1.0%	8.0%
75,000 and above	5.0%	1.0%	6.5%

In the web-fed example on the previous page, 35,000 copies of the book need 224,000 meters of paper. If you add a spoilage allowance of 7%, the total paper requirement comes to 239,680 meters. For the second book, the total paper requirement is 277,066 yards.

Paper Bulk (also known as calliper, volume, or thickness)
For a publisher it is useful to know how thick their book is going to be so that the designer can work out the correct spine width for the cover, the jacket, and any blocking on the spine of the case. It is possible to have a dummy made, but this is expensive. It is also useful when working out how many containers or trucks it will take to transport your books from the printer to the warehouse.

Bulk can be worked out either by measuring the thickness of a pile of sheets in fractions of an inch (or microns, at 1000 microns to a millimeter) and scaling up the result to equal the extent, or by using the manufacturer's volume basis for that paper.

If you are going to work out bulk using sheet thickness, it is better to measure a pile of sheets of the paper intended for the book, say ten or twelve, to take into account the air that occurs between the sheets in bound books but not with single sheets.

Say you are using an 80gsm paper, of which ten sheets bulk 1 millimeter): a 320-page book will consist of 160 leaves of this paper and bulk 16 millimeters.

Using the manufacturer's volume basis for that paper, the formula is:

$$\frac{weight\ in\ grams\ x\ volume\ basis\ x\ half\ the\ extent}{10,000}$$

A 320-page book, printed on an 80gsm Volume 14 paper will bulk 17.92 millimeters.

$$\frac{80gsm\ x\ Vol.14\ x\ 160 = 17.92mm}{10,000}$$

You can also work out the thickness (or calliper) of the paper by using the following formula:

$$\frac{gsm\ x\ volume\ basis}{10}$$

The calliper of the paper in the example above is 112 microns x 160 = 17,920 microns or 17.92mm.

These paper formulae are useful for giving you indicative figures. For absolute accuracy, you really should speak to your supplier.

If you are working in the USA, bulk is expressed in pages per inch (ppi). Most suppliers and paper merchants will provide you with tables showing you the ppi for various grades of paper in a range of weights. The table below acts as an indication of how bulk changes with weight and grade. For really accurate figures, you should speak to your supplier.

GRADE	50LB	55LB	60LB	65LB	70LB	75LB	80LB
Machine finished	490-560	440-510	410-470	380-440	35-400	330-370	300-350
Matt coated	620-700	570-630	520-560	480-520	440-470	400-430	370-390
Coated	800-850	730-790	640-740	590-680	530-620	490-580	440-530

Abbreviations

AFNOR	Association Française de Normalisation [French standards agency]
BIEC	Book Industry Environmental Council
CEO	chief executive officer
CEPI	Confederation of European Paper Industries
CoC	chain of custody
CRC	Carbon Reduction Commitment
CRT	cathode ray tube
Defra	Department for Environment, Food and Rural Affairs [UK]
EAG	Environmental Action Group [UK]
EMS	environmental management system
EPA	Environmental Protection Agency [US]
EPAT	environmental paper assessment tool
EPR	extended product responsibility
EU	European Union
EUTR	European Union Timber Regulation
EWP	European Water Partnership
EWS	European Water Stewardship
FLEGT	Forest Law Enforcement, Government, and Trade
FSC	Forest Stewardship Council
GHG	greenhouse gas
ICFPA	International Council of Forest and Paper Associations
IPCC	Intergovernmental Panel on Climate Change
ISO	International Organization for Standardization
LCA	Lifecycle Analysis

LCD	liquid crystal display [screen]
NGO	non-governmental organization
PEFC	Program for Endorsement of Forest Certification
PEST	political, economic, social, technological [factors for business]
PFCR	Product Footprint Category Rules
POD	print on demand
PREPS	Publishers' database for Responsible Environmental Paper Sourcing
REDD	Reducing Emissions from Deforestation and Forest Degradation
SME	small-to-medium size enterprise
STEEPLE	social, technological, economic, environmental, political, legal, ethical [factors for business]
STEP	political, economic, social, technological [factors for business]
UN	United Nations
VOCs	volatile organic compounds
VPA	voluntary partnership agreement
WBCSD	World Business Council for Sustainable Development
WRG	Water Resources Group
WRI	World Resource Institute
WWF	World Wide Fund for Nature [formerly World Wildlife Fund]

Bibliography

ARTICLES/REPORTS

Asner, Michael A. et al. "The Lacey Act Gives Gibson Guitar the Blues," *White Collar Crime Report*, 16 December 2011. http://www.illegal-logging.info/uploads/ArnoldPorterLLPBNAWhiteCollarCrimeReport121611.pdf

Ausli, Andrew et al. "Trees in the Greenhouse," *WRI*, June 2008. http://pdf.wri.org/trees_in_the_greenhouse.pdf

Baumert, Kevin A. et al. "Navigating the Numbers," *WRI*, December 2005. http://pdf.wri.org/navigating_numbers.pdf

Blue Plant Prize Laureates. "Environment and Development Challenges: The Imperative to Act," Blue Planet Prize, The Asahi Glass foundation, 2012. http://www.af-info.or.jp/en/bpplaureates/doc/2012jp_fp_en.pdf

Bristow, Rosie. "Talk Point: The Water-Energy-Food Nexus," *The Guardian*, 23 January 2012. http://www.guardian.co.uk/sustainable-business/talk-point-water-food-energy-nexus

Burrows, David. "Walking on Water," *The Guardian*, 20 May 2011. http://www.guardian.co.uk/sustainable-business/water-footprinting-local-issue

Carbon Trust. "Policy and Markets," 2012. http://www.carbontrust.com/client-services/advice/policy-markets

CEPI. "Carbon Footprint for Paper and Board Products: Appendices," 2007. http://62.102.106.97/Objects/1/Files/Carbon%20Footprint%20appendices.pdf

Defra. "Consumer Understanding of Green Terms: A Supplementary Report on Consumer Responses to Green Labels," 2011. http://randd.defra.gov.uk/Document.aspx?Document=Labellingfinalforpublication.pdf

Defra. "Consumer Understanding of Green Terms," 2011 http://randd.defra.gov.uk/Document.aspx?Document=EV0518_9994_FRP.pdf

Doughty Centre. "Corporate Responsibility Champions Network: A 'How To' Guide," Cranfield School of Management, 2009 http://www.som.cranfield.ac.uk/som/dinamic-content/research/doughty/CRChampions.pdf

E360 digest. "Nearly half of electricity at UK businesses wasted during off hours," *Yale Environment 360*, 7 February 2012 http://e360.yale.edu/digest/nearly_half_of_electricity_at_uk_businesses_wasted_during_off_hours/3321/

EIA. "Setting the Story Straight, The US Lacey Act: Separating Myth from Reality," 2010 http://www.eia-global.org/PDF/Report--Mythbusters--forest--Feb10.pdf

Engelman, Robert. "The Impact of Ecological Limits on Population Growth," *The Guardian*, 14 October 2011 http://www.guardian.co.uk/environment/2011/oct/14/1

Hohnen, Paul. "Sustainability reporting: immediate choices for the future," *The Guardian*, 30 January 2012 http://www.guardian.co.uk/sustainable-business/sustainability-reporting-integrated-business

Jackson, Tim (Prof.). *Prosperity without Growth? The Transition to a Sustainable Economy*. Sustainable Development Commission, March 2009

Jowit, Juliette. "Economic Report into Biodiversity Crisis Reveals Price of Consuming the Planet," *The Guardian*, 21 May 2010 http://www.guardian.co.uk/environment/2010/may/21/biodiversity-un-report

Lenton, Tim (Prof.) et al. "Major Tipping Points in the Earth's Climate System and Consequences for the Insurance Sector," WWF, 2009 http://assets.wwf.org.uk/downloads/tipping_point_report.pdf

Merkies, Judith and Lowitt, Eric. "Leasing Could Provide the Route to a Circular and Self-Sustaining Economy," *The Guardian*, 20 April 2012 http://www.guardian.co.uk/sustainable-business/leasing-route-circular-economy

Murray, James. "Google Tops Greenpeace IT Ranking," *The Guardian*, 8 February 2012 http://www.guardian.co.uk/environment/2012/feb/08/google-greenpeace-it-ranking

Nestlé. 26 January 2012 http://www.nestle.com/Media/NewsAndFeatures/Pages/Nestle-water-World-Economic-Forum.aspx

Pearce, Fred "Green Advertising Rules are Made to be Broken," *The Guardian*, 23 March 2010 http://www.guardian.co.uk/environment/2010/mar/23/green-claims

Purt, Jenny. "Discussion Round-Up: SMEs Engaging Employees in Carbon Reduction," *The Guardian*, 10 February 2012 http://www.guardian.co.uk/sustainable-business/carbon-reduction-employee-engagement-sme-discussion-round-up

Science Daily. "Land Use Change Influences Continental Water Cycle," 2011 http://www.sciencedaily.com/releases/2011/06/110628111842.htm

The Economist. "Following the Footprints," 2011 http://www.economist.com/node/18750670

Tregaskis, Shiona et al. "How the World Population Got to 7 Billion–Interactive," The Guardian, 28 October 2011 http://www.guardian.co.uk/environment/interactive/2011/oct/28/world-population-growth-7-billionth-person?INTCMP=SRCH

Uryu, Yumiko et al. "Deforestation, Forest Degradation, Biodiversity Loss and CO_2 Emissions in Riau, Sumatra, Indonesia," WWF, 27 February 2008 http://www.worldwildlife.org/who/media/press/2008/WWFBinaryitem7625.pdf

Vidal, John. "World Pays Ecuador not to Extract Oil from Rainforest," *The Guardian*, 30 December 2011 http://www.guardian.co.uk/environment/2011/dec/30/ecuador-paid-rainforest-oil-alliance

Waughray, Dominic. "Water at Davos 2012: Launching a New Model of Partnership," *The Guardian*, 26 January 2012 http://www.guardian.co.uk/sustainable-business/davos-water-scarcity

World Economic Forum. "Risks in Focus 3: The Water-Food-Energy Nexus," http://www.weforum.org/videos/risks-focus-3-water-food-energy-nexus

BOOKS

Publishers Association. *PA Statistics Yearbook 2011.* The Publishers Association, 2012

Swallow, Lisa. *Green Business Practice for Dummies.* Wiley, 2009

WEBSITES

Alliance for Water Stewardship
http://www.allianceforwaterstewardship.org/

Book Industry Environmental Council
http://www.bookcouncil.org/index.html

Carbon Disclosure Project
https://www.cdproject.net/en-US/Pages/HomePage.aspx

Carbon Reduction Commitment
http://www.carbonreductioncommitment.info/

CEO Water Mandate
http://ceowatermandate.org/

Environmental Paper Network's Paper Calculator
http://calculator.environmentalpaper.org/home

EPA on greenhouse gas reporting rules
http://www.epa.gov/climatechange/emissions/
ghgrulemaking.html

European Water Partnership
http://www.ewp.eu/

FLEGT Voluntary Partnership Agreements
http://www.euflegt.efi.int/portal/

GHG Protocol
http://www.ghgprotocol.org/

Green Press Initiative
http://www.greenpressinitiative.org/

Illegal Logging, a Chatham House managed site
http://www.illegal-logging.info/index.php

New Economics Forum on Well-Being
http://www.neweconomics.org/programmes/well-being

**Product Carbon Footprint (PCF) World Forum
on French environmental labeling**
http://www.pcf-world-forum.org/tag/grenelle-2/

REDD Monitor
http://www.redd-monitor.org/

UN REDD Program
http://www.un-redd.org/Home/tabid/565/Default.aspx

Water Footprint Network
http://www.waterfootprint.org/?page=files/home

WWF Check Your Paper
http://checkyourpaper.panda.org/

WWF on Deforestation
http://wwf.panda.org/about_our_earth/about_forests/
deforestation

Index

Acknowledgments

We would like to thank the many people who have contributed to this book. Particular thanks go to Mark Gough, who played such an important part in shaping the book and who has always shared his knowledge so generously. Thanks also to Liz Warman, Deborah Wright, Peter Hughes, and Shaun Hodgkinson at Penguin; to all who helped at Acona; and to Alison Kennedy at Egmont. And the biggest of thanks to our families, to Simon, Bryony, Stephen and to Berne, Hannah, and Lizzie for their incredible support and encouragement.

Every effort has been made to trace all the copyright holders, but if any have been inadvertently overlooked, the publishers will be pleased to make the necessary arrangements at the earliest opportunity.